The Moon Landing

Mankind's Greatest Adventure

The view of earth from *Apollo 11*, July 20, 1969.

LIFE

The Moon Landing

EDITORIAL DIRECTOR Kostya Kennedy
DIRECTOR OF PHOTOGRAPHY Christina Lieberman
CREATIVE DIRECTOR Gary Stewart
DESIGNER Patty Alvarez
WRITER Steve Rushin
COPY CHIEF Parlan McGaw
COPY EDITOR Joel Van Liew
PICTURE EDITOR Rachel Hatch
WRITER-REPORTER Ryan Hatch
PHOTO ASSISTANT Steph Durante
PREPRESS DESKTOP SPECIALIST Paige E. King
COLOR QUALITY ANALYST Jill M. Hundahl

MEREDITH SPECIAL INTEREST MEDIA
VICE PRESIDENT & GROUP PUBLISHER Scott Mortimer
VICE PRESIDENT, GROUP EDITORIAL DIRECTOR Stephen Orr
VICE PRESIDENT, MARKETING Jeremy Biloon
EXECUTIVE ACCOUNT DIRECTOR Doug Stark
DIRECTOR, BRAND MARKETING Jean Kennedy
SALES DIRECTOR Christi Crowley
ASSOCIATE DIRECTOR, BRAND MARKETING Bryan Christian
SENIOR BRAND MANAGER Katherine Barnet

EDITORIAL DIRECTOR Kostya Kennedy
CREATIVE DIRECTOR Gary Stewart
DIRECTOR OF PHOTOGRAPHY Christina Lieberman
EDITORIAL OPERATIONS DIRECTOR Jamie Roth Major
MANAGER, EDITORIAL OPERATIONS Gina Scauzillo

SPECIAL THANKS Brad Beatson, Melissa Frankenberry, Samantha Lebofsky, Kate Roncinske, Laura Villano

MEREDITH NATIONAL MEDIA GROUP
PRESIDENT Jon Werther
PRESIDENT, MEREDITH MAGAZINES Doug Olson
PRESIDENT, CONSUMER PRODUCTS Tom Witschi
PRESIDENT, CHIEF DIGITAL OFFICER Catherine Levene
CHIEF REVENUE OFFICER Michael Brownstein
CHIEF MARKETING & DATA OFFICER Alysia Borsa
MARKETING & INTEGRATED COMMUNICATIONS Nancy Weber

SENIOR VICE PRESIDENTS
CONSUMER REVENUE Andy Wilson
CORPORATE SALES Brian Kightlinger
DIRECT MEDIA Patti Follo
RESEARCH SOLUTIONS Britta Cleveland
STRATEGIC SOURCING, NEWSSTAND, PRODUCTION Chuck Howell
DIGITAL SALES Marla Newman
PRODUCT & TECHNOLOGY Justin Law

VICE PRESIDENTS
FINANCE Chris Susil
BUSINESS PLANNING & ANALYSIS Rob Silverstone
CONSUMER MARKETING Steve Crowe
SHOPPER MARKETING Carol Campbell
BRAND LICENSING Steve Grune

VICE PRESIDENT, GROUP EDITORIAL DIRECTOR Stephen Orr
DIRECTOR, EDITORIAL OPERATIONS & FINANCE Greg Kayko

MEREDITH CORPORATION
PRESIDENT & CHIEF EXECUTIVE OFFICER Tom Harty
CHIEF FINANCIAL OFFICER Joseph Ceryanec
CHIEF DEVELOPMENT OFFICER John Zieser
PRESIDENT, MEREDITH LOCAL MEDIA GROUP Patrick McCreery
SENIOR VICE PRESIDENT, HUMAN RESOURCES Dina Nathanson

CHAIRMAN Stephen M. Lacy
VICE CHAIRMAN Mell Meredith Frazier

Published by LIFE Books, an imprint of Meredith Corporation • 225 Liberty Street • New York, NY 10281

Vol. 19, No. 16 • June 21, 2019

CONTENTS

Edwin "Buzz" Aldrin on the Sea of Tranquility, July 1969.

FASCINATION WITH THE MOON IS as old as human civilization, captured in lore and scripture, art and film. Clockwise from top: An iconic image from the 1902 silent film *A Trip to the Moon*; an illustration (circa 1890) of Shakespeare's *Romeo and Juliet*; and, from 1958, an illustration in *La Domenica del Corriere*, an Italian weekly that told of mankind's desires to one day set foot on the faraway orb.

LA DOMENICA DEL CORRIERE

The End of the Impossible

On July 20, 1969, humankind's collective perspective changed forever

» BY STEVE RUSHIN

When Neil Armstrong emerged from the lunar landing module *Eagle* on July 21, 1969, the moon's gray-scale landscape was reflected in the gold-tinted visor of his helmet, which resembled the curved glass fronts of the television screens through which half a billion people—one sixth of the earth's population—witnessed the completion of *Apollo 11*'s journey from Cape Kennedy to the Sea of Tranquility. From the earth to the moon.

The sun was setting low in the lunar sky at 10:56 p.m. Eastern Daylight Time—the newspapers said the five-foot-11 Armstrong would cast a 35-foot shadow—and as he descended the nine-rung ladder of the *Eagle,* he held on to its side rails. When his oversize left moon boot made its impression on the desolate surface, the moon dust was as fine as talcum. "Like powdered charcoal," Armstrong would say of the footprint left by the footstep that was the most watched event in human history.

But what was under that powdered charcoal? "I was worried that the moon might be too soft and that he might sink in too deeply," said Viola Armstrong, his mother, who was watching at home in Wapakoneta, Ohio, where as a teenager her boy gazed into the night sky and dreamed of flight. On the moon, however, her son weighed one sixth of his earthly 165 pounds, and he didn't disappear into a porous surface—thus instantly dispelling ancient myths that the moon was made of something else, like green cheese.

"That's one small step for man," Armstrong said. "One giant leap for mankind." What he intended to say, of course, was: "That's one small step for a man, one giant leap for mankind." But the indefinite article *a* was dropped, leaving a redundant phrase that instantly entered the lexicon anyway.

Shakespeare portrayed the moon as unfaithful and sinister. He called the waxing and waning moon an "arrant thief," "the fickle moon, the inconstant moon." It is she who "makes men mad," the Bard wrote, but the world was moonstruck long before Romeo and Juliet and Othello. *Lunacy* and *lunatic* and *lunar* all derive from the same root, the name *Luna,* the ancient Roman goddess of the moon. From werewolves to legends of the man in the moon, this pale orb—governess of tides, mythical synchronizer of menstrual cycles—has always been a source of mystery, fascination, and dread.

Jules Verne, in his 1865 novel, *From the Earth to the Moon,* wrote about three men flying to the moon in a small capsule: "In spite of the opinions of certain narrow-minded people, who would shut up the human race upon this globe... we shall one day travel to the moon, the planets, and the stars with the same facility, and rapidity, and certainty as now make the voyage from Liverpool to New York." And Frank Sinatra expressed this universal longing in a song played by the *Apollo 10* astronauts during their 1968 dress rehearsal to lunar landing: "Fly me to the moon and let me play among the stars."

The moon made bards of the unlikeliest poets. You didn't have to be Cole Porter ("a trip to the moon on gossamer wings") for the moon to be your muse. Stepping onto its surface 19 minutes after Armstrong did—with Michael Collins orbiting the moon alone in the command module *Columbia,* awaiting rendezvous with the *Eagle*—Buzz Aldrin marveled out loud at the "magnificent desolation" of the lunar landscape. And another enduring phrase was minted.

When in the run-up to the *Apollo 11* mission Aldrin was asked by LIFE why he should go to the moon at all, he replied: "If you are a fighter pilot, you want to get hold of the hottest thing you can. And having flown that, you ask yourself: 'What else can I fly?' You come down to the ultimate, the space program. It's there."

The answer echoed what mountaineer George Mallory said in 1924 when asked why he climbed Mount Everest: "Because it's there." President John F. Kennedy addressed the very question when he spoke in Houston on September 12, 1962, ostensibly to open NASA's Manned Spaceflight Center there: "But why, some say, the moon?" the President told an audience at Rice University. "Why choose this as our goal? And they may well ask why climb the highest mountain? Why, 35 years ago, fly the Atlantic? Why does Rice play Texas?"

One might now ask, from the remove of 50 years: What was the big deal? Why was the world so moonstruck in the summer of 1969? Why did the top records on the charts on July 19 include such entries as "In the Year 2525" (No. 1, with its futuristic musings on starlight), "Good Morning Starshine" (No. 3), and "Bad Moon Rising" (No. 11)? David Bowie's "Space Oddity" had just been released eight days earlier, about a man "sitting in a tin can, far above the world." The president of the Toy Manufacturers of America, Lionel Weintraub, said that summer that the *Apollo 11* astronauts had "practically turned the toy industry into a branch of the National Aeronautics and Space Administration." Tang, the powdered orange breakfast drink imbibed by astronauts in space since 1962, had become an unlikely breakfast staple for kids. "Tang sucks," Buzz Aldrin would say in 2013, but in 1969 anything touched by the moon was cool.

Pan Am had 24,000 passengers wait-listed for its first commercial trip to the moon, though that expedition—called Moonflite—was still very much on the drawing board. TWA took 4,000 requests for their own hypothetical long-haul flight to the moon. The hotel magnate Barron Hilton described his vision for a Lunar Hilton. Rental car industry executives envisioned a central car-rental hub in the moon's Copernicus Crater, with drives recommended to the Sea of Nectar and the Bay of Dew. Corporate titans were measuring the moon for drapes in the summer of 1969. Mankind's move-in date was imminent.

SPACE MANIA WAS IN FULL THROTTLE IN THE 1960s as the mission crept near. Clockwise from top left: On Broadway, the musical *Hair* proclaimed the "Age of Aquarius"; Stanley Kubrick's film *2001: A Space Odyssey* inspired items like this Lunar Hilton room key and reservation form; English rocker David Bowie released his song "Space Oddity"; and a passenger secured a card promising a seat on a future Pan Am moon flight.

EIGHT YEARS BEFORE liftoff, space suits were tested in the Mojave Desert (above), which had a surface, NASA said, not unlike the moon's. Across the Atlantic in France, a different sort of moon suit was donned in 1967, this one (far right) designed by Paco Rabanne and made of flexible metal. Kids, of course, were awestruck by the moon, and toys such as this *Apollo 11 Eagle* lunar module were shipped around the world.

Of course, the moon landing was much more than a show of crass commercial imperialism, though it was clearly that: Car washes, mattress stores, and used-car dealers offered moon landing specials. Walgreens had "the lowest prices this side of the moon."

And the moon landing was more than an expression of national pride, though it was clearly that as well. ("The Moon Is Made of American Cheese" proclaimed one triumphant T-shirt.) The moon landing inspired global awe and admiration, with parochial outbreaks of envy. Some 3,500 credentialed members of the worldwide press carried the launch of *Apollo 11* live in 33 languages to people around the globe. Every nation on earth watched or—in certain instances—made an ostentatious show of not watching. From Castel Gandolfo, his retreat south of Rome, Pope Paul VI gazed at the lunar landing area through a telescope—336 years after the Roman Catholic church had declared Galileo a heretic for saying the earth revolves around the sun and not vice versa.

Death row inmates at Sing Sing prison, ordinarily denied television, were given an exception to watch the landing. Fire trucks in Guayaquil, Ecuador, sounded their horns in celebration of those first steps. A baby born in Beirut was named Apollo. A newborn in Scotland was named for all three astronauts: Neil Edwin Michael Robertson. And outside Toledo, Ohio, not far from Armstrong's hometown, Mrs. Delmar Moon gave birth at 12:03 p.m. on Monday, July 21, to the insuperably named Neil Armstrong Moon.

The citizens of Robbinsdale, Minnesota, named their newest temple of education Neil A. Armstrong High School, an extraordinary honor for a living 38-year-old, with few previous parallels—among them Armstrong's boyhood hero, Charles Lindbergh, who shrunk the world in the way that Armstrong was now shrinking the solar system. Crowds celebrated in Trafalgar Square in London and on the Champs-Élysées in Paris. The astronauts' achievement reportedly went unmentioned in Chinese media and was downplayed in a 52-word dispatch from the Soviet news service TASS. Elsewhere behind the Iron Curtain, a teenager in Yugoslavia suggested that the moon landing had rendered romance dead. "Now the moon is real," he lamented to a reporter, "and lovers won't have it for themselves alone anymore."

The experience of the Apollo astronauts, looking back on a borderless blue marble—in what would become a famous photograph, "Earthrise"—with its unseen wars in Vietnam and the Middle East, literally brought a new perspective to the citizens of earth. Many aspired for the first time to better stewardship of the planet (the first Earth Day was held the following April) and demanded an end to global hostilities. There was a realization—however fleeting it might prove—that we are "on a mote of dust suspended in a sunbeam," as Carl Sagan would write years later. After landing on the moon but before exiting the *Eagle*, Buzz Aldrin, an elder in his Presbyterian church, took Holy Communion—bread and wine, drunk from a chalice—that he brought with him in a personal carry-on bag. He didn't broadcast the fact over the radio, for the moon landing was to be an ecumenical experience, for all people, of any religion or none at all.

And while Armstrong and Aldrin assembled and planted a wire-stiffened $5 American flag meant to ripple forever in the vacuum of space, many in the federal government argued against that show of American imperialism. Congress eventually ordered the flag to be planted. In addition, a stainless-steel plaque affixed to the ladder of the *Eagle* that remains on the moon to this day reads: "Here men from the planet Earth first set foot upon the Moon, July 1969, A.D. We came in peace for all mankind." The text NASA submitted to the White House had read "We come in peace for all mankind," but Nixon speechwriter William Safire thought that sounded "like the sort of thing you'd say to Hollywood Indians," so he changed the verb tense.

These prosaic concerns aside, the moon landing more than anything else instantly augured a universe of infinite possibility. Earthbound human beings looked at Armstrong and Aldrin's convex sun visors and saw themselves reflected back, but in a new and gold-tinted light. "If they can put a man on the moon, surely they can . . ." went the new phrase. And the end of that sentence was some version of "do anything." Cure cancer. Stop famine. End war.

The previous exclamation of boundless opportunity—"the sky's the limit"—was rendered instantly inadequate. Armstrong was still 48 hours from setting foot on the moon when Mission Control radioed the astronauts on July 17, 1969, with the day's news. That digest included U.S. Vice President Spiro Agnew's announcement to the world that America intended to put a man on Mars by the futuristic year of 2000. Speeding toward the moon on its 238,000-mile journey, borne aloft on a column of fire in Florida the day before, *Apollo 11* carried three astronauts spiritually descended from the Argonauts—explorers on an ancient journey of discovery. They were, as Armstrong would tell Congress in 2010, "learning to sail upon this new ocean."

The mark these men left remains indelible 50 years later, as does Armstrong's first step. It will be ever thus. "The first footprints on the Moon will be there for a million years," as NASA notes. "There is no wind to blow them away." ●

Before the Moon
A Story in Pictures

THE MERCURY SEVEN

PROJECT MERCURY, THE FIRST U.S. HUMAN SPACEFLIGHT program, ran from 1958 to 1963. The Mercury flights essentially functioned as tests for the Apollo missions later in the decade, while also alerting Russia—the rival nation was first, by a few weeks, to put a man into space—that the United States was serious about getting to the moon. Above, the Mercury astronauts: (top row, from left) Alan Shepard, Virgil Grissom, Leroy Cooper, (bottom row, from left) Walter Schirra, Donald Slayton, John Glenn, Malcolm Carpenter.

MAY 5, 1961

GOING SUBORBITAL

The first American in space was Alan Shepard (here, heading to the launchpad), when he completed a 15-minute suborbital flight on May 5, 1961.

FEBRUARY 20, 1962

DOWN THE HATCH

John Glenn (right) became the first American to orbit earth, doing so three times in less than five hours on February 20, 1962, aboard the *Friendship 7* module on the Mercury Titan rocket. A World War II veteran and later a United States senator from Ohio, Glenn was awarded the Presidential Medal of Freedom in 2012. He died in 2016 at age 95.

hip

JUNE 3, 1965

HANGING OUT

Gemini 4 astronaut Ed White made the first space walk, or EVA (extravehicular activity), spending more than 20 minutes outside the spacecraft connected by a tether that supplied oxygen. The space walk was in many ways a precursor to the coming moon walk, testing man and equipment in the vacuum of space. White used a "zip gun" to propel himself in the weightless void of outer space. Commanded by James McDivitt, *Gemini 4* made 62 trips around the earth.

JULY 16, 1969

MOON BOUND

The Saturn V rocket created decibel levels of 204 and traveled up to 25,000 miles per hour—in pursuit of one of mankind's most audacious dreams.

"Is your son going to fly out into space?"

From all of humanity, three men—fit, tested, and resolute—were chosen to make our most magnificent leap

ONCE THE REALM OF SCIENCE fiction, the idea of landing on the moon became reality in July 1969 when Neil Armstrong, Buzz Aldrin, and Michael Collins made it there and back, capturing this photo from 10,000 miles out in space, as *Apollo 11* returned to earth.

ON MAY 25, 1961, PRESIDENT John. F. Kennedy addressed a joint session of Congress to seek legislation for the national goal of reaching the moon "before this decade is out," saying that "no single space project in this period will be more impressive to mankind." Kennedy did not live to see the moon landing. He was assassinated in Dallas in 1963.

In an address to the assembled members of Congress on May 25, 1961, President John F. Kennedy declared, "I believe that this nation should commit itself to achieving the goal, before this decade is out, of landing a man on the moon and returning him safely to the earth."

Six weeks earlier, Yuri Gagarin of the Soviet Union had become the first human to travel in space, completing one orbit of the earth in a capsule called *Vostok*. Five days after that, the United States invaded the Bay of Pigs in Cuba, failing in a spectacular way to overthrow Soviet-backed dictator Fidel Castro. The cold war was starting to warm up—and also shifting to an extraterrestrial setting. The moon "was no longer the governess of the tides, the lovers' beacon, the celebrated mistress of song," as Andrew Chaikin put it in his history of the Apollo program, *A Man on the Moon*. "It was a target, a cold war beachhead in the sky. It was NASA's moon."

With a speech at Rice University the following year, Kennedy affirmed this, and began the countdown to a more historic countdown, little more than eight years later, at what would be

They had the "right stuff": an indefinable courage and savoir faire, an immunity to danger and to the perils of fame.

called Cape Kennedy in central Florida. The Apollo program would be the third NASA project to send men into space. Project Mercury, begun in 1959, was the first. Its mission was to send a manned spacecraft into orbit around the earth, and to return both man and spacecraft safely back to terra firma. Before that happened, NASA attempted two suborbital solo missions. Alan Shepard was delivered 116 miles above Cape Canaveral on a Mercury-Redstone rocket on May 5, 1961, and returned intact 15 minutes later—and thus a phrase would forever attach itself to his name: "the first American in space." He was quickly followed by Gus Grissom, "the second American in space."

On February 20, 1962, John Glenn made the first of Mercury's four orbital missions around the earth and was lionized, instantly, as an American hero, a latter-day Lindbergh. "The astronauts, all of us, really believed we were locked in a battle of democracy versus communism," Glenn would say later, "where the winner would dominate the world."

The men in the program—along with Shepard, Grissom, and Glenn were Scott Carpenter, Wally Schirra, Gordon Cooper, and Deke Slayton—became renowned as the Mercury Seven. The Mercury Seven were said

CHOSEN AS THE THREE MEN best (ahem) suited for the job, astronauts (from left) Neil Armstrong, Michael Collins, and Buzz Aldrin posed for an official portrait two months before their historic flight.

to possess the "right stuff," a phrase coined by New Journalism pioneer Tom Wolfe, who wrote a nonfiction classic of that name about them. The right stuff was an indefinable courage, humility, and savoir faire, an immunity to both danger and the perils of fame. NASA required men who possessed this stuff (while also being no taller than five foot 11) for future space travel.

In 1961, Project Gemini began to lay the groundwork—the spacework—for the project to follow. Where the Mercury flights were solo, Gemini (Latin for "twins") flights were two-man missions. They were powered by a Titan II rocket and acted as a bridge of sorts to the three-man Apollo missions charged with putting a man on the moon. On June 3, 1965, astronaut Ed White opened the hatch of *Gemini 4* and stepped into space, his white boots silhouetted by the blue Pacific Ocean surrounding Hawaii. A long tether acted as a galactic umbilical cord by which White returned to the capsule after 23 minutes, though by then he was over the Gulf of Mexico. White had become the first American to walk in space, and in doing so he blazed a kind of vapor trail for the ultimate space walk to follow, on the moon, four years later: one small step before Armstrong's small step.

It was Abe Silverstein, director of NASA's Office of Space Flight programs, who suggested the name Apollo, in consonance with the ancient names of Gemini and Mercury. In Greek mythology, Apollo was, among many other things, a fearsome archer, capable of hitting a distant target. And now one of NASA's principal tasks was to find men willing to sit athwart that arrow as it traveled seven miles per second in an effort to escape the earth's gravitational pull on its 238,000-mile voyage to the moon.

ON BREAK FROM FLIGHT training in Florida, Armstrong visited his grandmother (top) while later, in 1962, his parents appeared on the CBS show *I've Got a Secret*, announcing to the country before NASA could their son's new gig that, soon, had him rocketing at Mach 3 after a reentry from 207,000 feet during simulation training. After a stressful landing, he was all smiles, opposite.

NASA
BEWARE
FORCE X15
NO.56-6670
NASA

Neil Armstrong moved 16 times by the time he was 14 years old, each move within his home state of Ohio, for whom his father, Stephen, was a government auditor. By the time the family settled in Wapakoneta, where Neil attended Blume High School, the boy was building model airplanes and suspending them above his bed. Was it any wonder, beneath these models, that he had a recurring dream of flight—of hovering in the air?

In 1947, 16-year-old Neil Armstrong peered through the eight-inch telescope belonging to his neighbor Jacob Zint, a draftsman who had built atop his garage an observatory that rotated on roller skate wheels. "At first Neil was too shy to ask me if he could look through my telescope and his mother would sometimes call up and smooth the way before he came over," Zint told the Associated Press in January of 1969. "But when he was here he was extremely interested. And you know, he talked more about the moon than any other part of the heavens, even then."

This story, published in papers throughout America in 1969, was a wild embellishment, Armstrong later told his biographer, James R. Hansen, author of *First Man*. Armstrong visited Zint's observatory once, to get his Boy Scout merit badge in astronomy, and evinced no particular fascination with the moon. But Zint told the story to so many reporters on the eve of the *Apollo 11* launch that it became part of the Armstrong lore, a moonstruck backstory, that endeavored to make the moon Armstrong's fate and part of America's manifest destiny.

The truth was no less interesting. Armstrong took his first flight at age six, when he and his father stopped at an airfield offering airplane rides on the way to Sunday school in Warren,

Armstrong got his pilot's license before his driver's license, already slipping the surly bonds of earth at age 16.

A YEAR BEFORE ENROLLING at West Point, Aldrin starred on Montclair High School's football team (above, Buzz as a senior, in 1947) and won a New Jersey state championship. Right: His athleticism came in handy years later during flight training as he (center, arms outstretched) and Gemini astronauts Charles Bassett (above right of Aldrin) and Theodore Freeman (lower right) floated weightless during a training exercise in a C-135 cargo plane in 1964, while a technician looked on.

DO NOT PAINT

SOMETIMES FORGOTTEN because he did not walk on the moon, astronaut Michael Collins (opposite) served as command module pilot and experienced perhaps the most daring aspect of *Apollo 11*: flying solo on the dark side of the moon. "I am alone now, truly alone, and absolutely isolated from any known life. I am it," he reported to Houston. Above: Class III of NASA's Aerospace Research Pilot School in 1962. Back row, from left: Al Atwell, Neil Garland, Jim Roman, Al Uhalt, and Joe Engle. Front row: Ed Givens, Tommie Benefield, Charlie Bassett, Greg Neubeck, and Collins.

Ohio. Enthralled, Neil took flying lessons. In Wapakoneta, 59 miles straight north of Orville and Wilbur Wright's first bicycle shop in Dayton, Armstrong got his pilot's license before his driver's license, already slipping the surly bonds of earth at age 16.

He wasn't an athlete. (Even as an astronaut, smoking a monthly cigar, Armstrong was not the fitness fanatic that some of his colleagues were. "I believe that every human being has a finite number of heartbeats available to him," he told LIFE in 1969, "and I don't intend to waste any of mine running around doing exercises.") In fact, he was a bit of a nerd, interested in engineering, marching band, and his studies. He graduated from Blume at 16—"He thinks, he acts, 'tis done," read the quote next to his yearbook photo—and began to study engineering at Purdue University, in West Lafayette, Indiana, where he met Janet Shearon, a home economics major from suburban Chicago.

Suspending his studies at Purdue to serve in the Korean War, Armstrong flew 78 combat missions; landed a damaged jet on the deck of the aircraft

THREE MONTHS BEFORE liftoff, Collins (opposite) sat in a gondola during centrifugal training in Houston, part of hundreds of hours of prep for the voyage. The planning was comprehensive—14 years earlier, in Johnsville, Pennsylvania, the centrifuge spun researchers in circles to study the effects of gravity, or lack thereof.

carrier *Essex*; and was shot down behind enemy lines and rescued the same day. After returning to and graduating from Purdue, Armstrong became a civilian test pilot at Edwards Air Force Base in California, where he worked for seven years, from 1955 to 1962, testing the rocket-powered X-15 while wearing the kind of silver flight suit and windowed helmet that inspired *Plan 9 from Outer Space* and other drive-in movies of the era.

"I have been in relatively high-risk businesses all of my adult life," he told Dodie Hamblin of LIFE, who wrote about the Armstrong family in 1969. "I suspect that on a risk-gain ratio, space missions would compare very, very favorably with those to which I have been accustomed in the past 20 years."

On January 28, 1962, Neil and Janet Armstrong suffered the loss of their two-year-old daughter, Karen Anne, to an inoperable brain tumor. Three weeks later, to worldwide attention, John Glenn made his orbit of the earth. "Was there an intensely personal—most likely, subconscious—relationship between Karen's death... and Neil's decision to submit his name for astronaut selection just a few months later?" asked Hansen in *First Man*. To which Armstrong's sister, June, replied: "The death of his little girl caused him to invest those

CONTINUED ON PAGE 35

"I suspect that on a risk-gain ratio, space missions would compare very, very favorably with those to which I have been accustomed in the past 20 years."

—NEIL ARMSTRONG

EXTENSIVE TRAINING took place inside the Bell Moon Landing Research Vehicle (here, in 1964), an instrument designed to simulate approaching the moon. Said Armstrong later, "What the LLTV gave you was not so much the seat-of-the-pants dynamics as the real-world visual . . . That and the fact that, if you make a mistake, you can't hit the reset button."

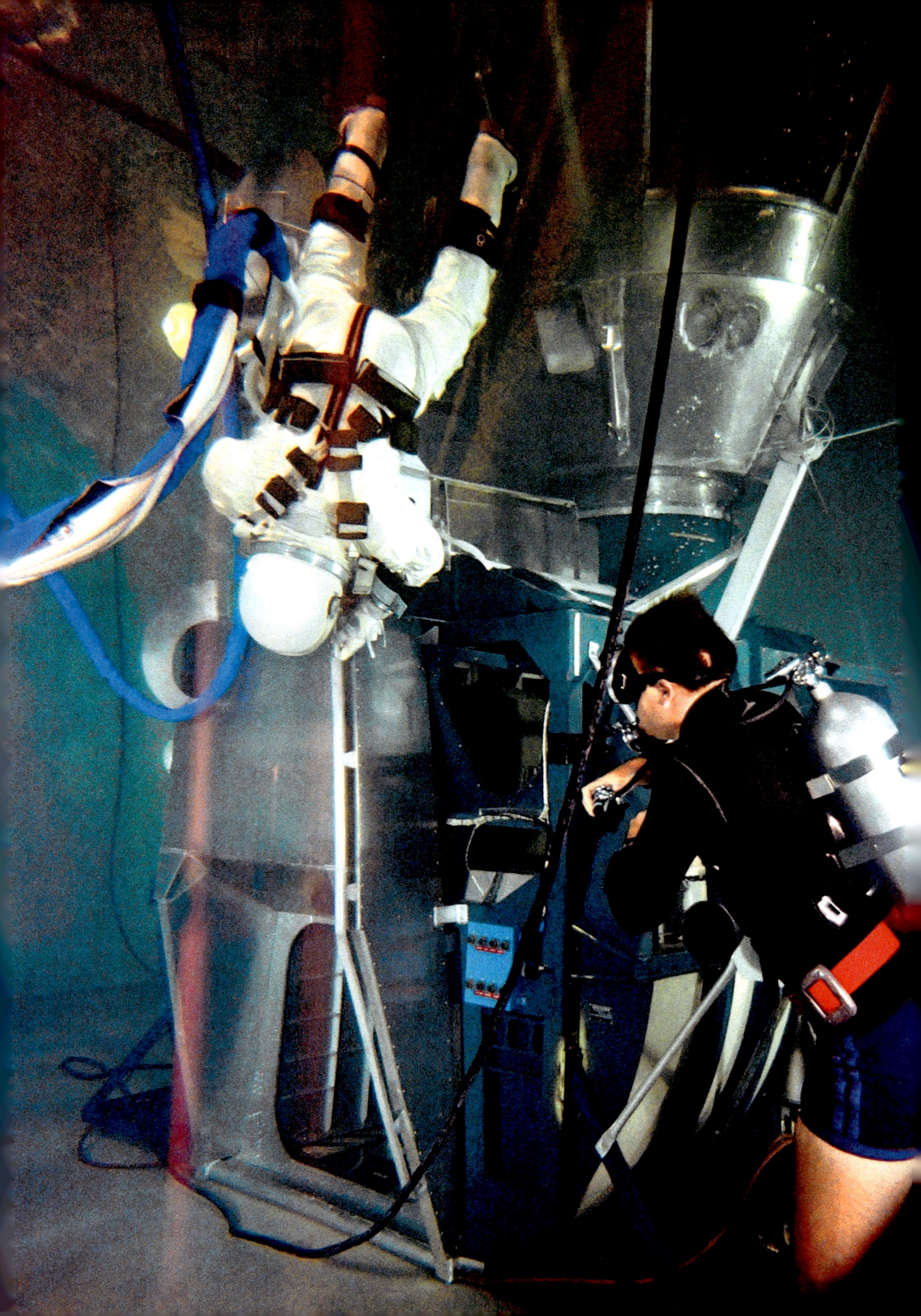

TO BEST IMITATE LIFE IN space, NASA trained its astronauts underwater to simulate zero gravity (opposite). In 1968, Armstrong narrowly escaped disaster when piloting the LLRV-1 at Ellington Air Force Base, ejecting to safety moments before impact.

CONTINUED FROM PAGE 31

energies into something very positive, and that's when he started into the space program."

On September 17, 1962, Stephen and Viola Armstrong appeared on the television quiz show *I've Got a Secret.* Celebrity panelist Betsy Palmer finally guessed their secret in a 20-question format: "Does [your secret] have anything to do with aeronautics?" she said. "Space? Is your son going to fly out into space soon? Is he a new astronaut?" "That's it," replied host Garry Moore. "Mr. and Mrs. Armstrong are very happy and a very proud couple. They're still kind of in a state of shock. For this afternoon at three p.m., their son, Neil, was named one of America's new astronauts."

In fact, he had become a civilian research pilot for NASA and the highest paid astronaut in the short history of that occupation, with an annual salary of $27,000. So began the high and the low of Armstrong's new existence—the space career unbound by gravity, and the gravitational pull of public life. Neil Armstrong, newly minted as an astronaut, learned of his selection into the astronaut program in a phone call from Deke Slayton. But America learned of it—indeed, first heard the name Neil Armstrong—on a CBS game show.

Armstrong was not going into space alone. He'd be joined by Edwin "Buzz" Aldrin Jr., the son of an Air Force colonel who had known both Orville Wright and engineer Robert Goddard. Aldrin Sr. knew Charles Lindbergh well enough to call him "Slim." It was Edwin Aldrin Sr. who introduced the *Spirt of St. Louis* pilot to Goddard, the father of modern rocket propulsion. The elder Aldrin would still maintain his pilot's license at age 73, by which time Buzz—his little sister pronounced "brother" as "buzzer," and the name stuck—was flying to the moon. Moon, incidentally, was the maiden name of Buzz Aldrin's mother. She was Marion Moon.

Buzzer graduated third in his class from West Point and joined the Air Force, flying 66 combat missions in Korea. He married Joan Archer (archer, like Apollo), piloted F-100s in Germany, and got his doctorate at the Massachusetts Institute of Technology, skipping over a master's degree. His first application to NASA's astronaut program was rejected before he was selected in 1963 for the Apollo program, in which he performed the most successful space walk to date, working outside *Gemini 12* for two hours and 18 minutes. While taking a photograph of earth, he told Houston, "Okay, tell everybody down there to smile."

The third member of the *Apollo 11* crew, Michael Collins, was born in Rome, where his father was stationed as an Army major. Raised on various military posts, Collins considered no single place on earth his home. He attended West Point, like Aldrin, and met his future wife, Patricia Finnegan, on an air base in France. The astronaut's wives would become a subplot to the story of *Apollo 11,* the beat of journalists reporting on "the home front." America had a bottomless curiosity about the astronauts' wives and how they kept the children occupied in their suburban Houston homes, while keeping their own worries in check as the men of the house went to the moon.

Collins, who by 1969 would have three children of his own—Kate, 10; Ann, seven; and Mike, six; to say nothing of their German shepherd,

ARMSTRONG AND ALDRIN spent nearly every waking hour training for *Apollo 11*'s moon mission. Above: In 1968, Aldrin strapped himself into a parachute harness at Perrin Air Force Base in Sherman, Texas. Opposite: In April 1969, Armstrong and Aldrin puttered around in Houston's Manned Spacecraft Center, placing lunar samples on a staged moon surface.

Dubhe—had applied to be an astronaut after Glenn circled the earth in the Mercury *Atlas 6* in 1962. On *Apollo 11*, he was to remain in the command module, *Columbia*, orbiting the moon, while Aldrin and Armstrong descended in the lunar module, *Eagle*, to walk on its surface.

"I think man has always gone where he could, he has always been an explorer," Collins told LIFE before liftoff. "There's a fascination in exploring and thrusting out to new places, but I can't say there is anything special about me—that I'm a natural explorer or I've always been bent in this direction—because it isn't true. I can be a good explorer, but so could other people. I really think the key is that man has always gone where he could and he must continue. He would lose something terribly important by having that option and not taking it."

In taking that option, Apollo astronauts were subjected to mental, physical, and psychological exercises designed to keep them alive on their voyage. Most notorious among these trials was the Johnsville human centrifuge, a sphere that required just seven seconds to get spinning at 175 miles an hour, creating g-force tests that Collins called "diabolical."

Apollo astronauts also endured jungle, mountain, and desert training to prepare for every eventuality, including sea survival should a capsule miss its splashdown target. For the splashdown itself, the *Apollo 11* astronauts rehearsed disembarking their capsule in swimming pools, and swimming underwater in space suits. On May 6, 1968, Armstrong was piloting the temperamental Lunar Landing Research Vehicle (LLRV) at

UNITED
STATES

Ellington Air Force Base, practicing the powered descent he would have to make to the moon's surface in the *Eagle* after detaching from the mother ship, *Columbia.* The controls on the LLRV failed while Armstrong was airborne and he ejected seconds before the vehicle crashed and exploded. Armstrong cut an almost serene figure floating to safety, his parachute deployed while an orange fireball consumed the spacecraft. It was a stark reminder, as if any were needed, that the moon landing remained a perilous ambition, hubristic, with echoes of the Icarus myth. But it was also a reminder that Armstrong—who retreated to his office and went about his day immediately after the accident—was exceedingly calm in the face of imminent disaster.

The superficial calm maintained by NASA was in part a display of sangfroid—the right stuff. But it was also a facade, the calm exuded by an airline pilot in turbulence. Unknown to the public, Aldrin's mother committed suicide in 1968, the year before he was to become a global phenomenon. By January of 1969, Joan Aldrin sensed that her husband was more solitary than usual. On January 9 of that year, the world learned why: NASA announced the three-man crew for *Apollo 11.* With Armstrong and Collins, Aldrin had been selected for the moon landing mission. Perhaps thinking of their three kids—Michael, 13, Janice, 12, and Andrew, 11—Joan Aldrin suddenly regretted her husband's profession. "I wished Buzz were a carpenter, a truck driver, a scientist," Joan said in the summer of 1969. "Anything but what he is."

Apollo 8 astronauts James Lovell and William Anders—who a month

WITH LAUNCH FAST approaching, Aldrin, Armstrong, and Collins (from left) continued their practice inside the Apollo "Boilerplate 1102" training capsule (behind them) in the Gulf of Mexico to undergo water egress training for their return to earth.

The three men were placed in the company of the great explorers—notwithstanding the debate about who would be the first on the moon.

earlier had made the first orbital flights around the moon—were named to the backup crew along with "rookie astronaut" Fred Haise. Wire services carried the biographies of all these men around the world. And while Collins was identified as the one who would remain in the *Columbia* as it orbited the moon, and Armstrong and Aldrin would be the lunar landers, "it will take months of practice in the make-believe missions," UPI reported, "to work out which man is first."

All three men were placed in the company of the great explorers of history—notwithstanding this new debate about which man would be the first on the moon. That person would be afforded the stature of Christopher Columbus, a new-world colossus. Indeed, not landing on the moon, Collins said, given Apollo's preparatory missions to get so close, "would be the equivalent of Columbus sailing within 60 miles of the coast of the New World and then turning back toward Queen Isabella and going home." For these men, there would literally be no turning back.

If NASA precedent held, Aldrin would be the first man on the moon. Armstrong had been named commander, and the commander on

Gemini missions always stayed in the capsule while his pilot—Aldrin, on the *Apollo 11* lunar module (or LM, pronounced "lem")—performed any extravehicular activities. Even Aldrin's father, Stephen, assumed that Aldrin would be first man. Yet the practical matter of the seating arrangement inside the LM dictated that Armstrong—nearer to the hatch opening than Aldrin, who was required to be in the pilot's seat—go first. "It was not something that I thought was really very important," Armstrong told biographer Hansen. "It has always been surprising to me that there was such an intense public interest about stepping onto the lunar surface, let alone who did it first."

Yet the public interest in the matter was intense. On April 14, 1969—on the day Katharine Hepburn and Barbra Streisand tied for best actress at the Oscars, and jury deliberations began in the trial of Sirhan Sirhan for the assassination of Robert Kennedy, and the 24th plane of the still-young year was hijacked to Cuba, and the new President, Richard Nixon, presented his domestic program to Congress—front pages across America quoted George Low, NASA manager of the Apollo spacecraft office, making the historic announcement: Should everything go according to plan, Neil Armstrong will become the first man to walk on the moon.

Toward that end, NASA took every possible precaution. Nixon proposed to have dinner with the astronauts at Cape Kennedy the night before launch, but NASA rebuffed the President, lest a Nixonian cough or sneeze compromise what novelist Norman Mailer called "the greatest adventure of man." ●

FRESH OUT OF THE BOX, ***Apollo 11*****'s S-IC booster for its Saturn V rocket (opposite) was erected in the spaceport's Vehicle Assembly Building. On the night before launch (above), Cape Kennedy sat quiet, waiting for the next day's affair.**

Margaret Hamilton in a mock-up of an Apollo command module.

"THERE WAS NO CHOICE BUT TO BE PIONEERS"

A woman in the mostly male field of software engineering, Margaret Hamilton wrote the code that landed the astronauts on the moon

NASA awarded its first major contract for the Apollo program on August 10, 1961. It went to the Instrumentation Lab at the Massachusetts Institute of Technology, where 24-year-old computer scientist Margaret Hamilton was working as a programmer, helping to pay the bills while her husband, James Cox Hamilton, attended Harvard Law School.

Hamilton was not only a working mother (the couple's young daughter, Lauren, sometimes slept in the lab at MIT), she was also a woman working among mostly male colleagues in an industry—computers—that was new. In that role, she was trying to achieve the unprecedented goal of putting a human being on the moon. "There was no choice but to be pioneers," she would later tell an audience at her alma mater, Earlham College, in her home state of Indiana. And so—in all of these areas—she set about pioneering.

MIT's contract was to provide the onboard navigation and guidance for the Apollo command and lunar modules. On *Apollo 11,* that meant *Columbia* and *Eagle,* with 500 million pairs of eyes upon them in July of 1969. Hamilton wrote the code for the computer software that Armstrong, Aldrin, and Collins used on board their spacecraft. In a famous photograph, Hamilton stands smiling next to a stack of computer code. The stack of papers is as tall as she is.

To describe the work she was doing—and to give it a prestige commensurate with other jobs in science—Hamilton coined the phrase "software engineering." That software was critical to the entire moon landing expedition, but especially so in the fraught seconds before Armstrong could finally declare, "The *Eagle* has landed."

Three minutes before setting down on the lunar surface, the *Eagle*'s "1202" and "1201" alarms sounded. Armstrong and Aldrin remained calm and conferred with Houston. The audience on earth didn't know that each alarm triggered an instant rebooting of the onboard computer, or that Hamilton's code prioritized the most important tasks at hand and overrode the less urgent ones. It was this code that gave the astronauts and their interlocutors in Houston the confidence not to abort the landing.

"Give us a reading on the 1202 program alarm," Armstrong said while descending to the moon's surface.

"We're go on that alarm," replied Charlie Duke in Houston.

Four months after that historic landing, Hamilton would be photographed in a mock-up of *Apollo 12,* which Pete Conrad—in mankind's second lunar landing—had just set down on a spot of moon known as Pete's Parking Lot. That NASA could make a landing of such precision, a quarter of a million miles from earth, was testament in large part to Hamilton and her colleagues. As the caption on the photograph of her distributed by the Associated Press read: "She headed the group that programmed *Intrepid*'s pinpoint landing on the moon."

After the Apollo program, Hamilton would work on SkyLab (the space station that orbited the earth for six years in the 1970s) and founded her own company. In her eighties, she remained the CEO of Hamilton Technologies, centered on the Universal Systems Language that she invented. The photograph of her standing next to that pile of code was recreated in three dimensions by Lego for a special Women of NASA tableau, for Hamilton has become, in the 21st century, a heroine to STEM (science, technology, engineering, and math) students and more recently to the public at large. In 2016—along with Robert De Niro, Bill Gates, Bruce Springsteen, and Michael Jordan, among others—Hamilton was awarded the Presidential Medal of Freedom by President Barack Obama.

"She symbolizes that generation of unsung women who helped send humankind into space," Obama said. "Her software architecture echoes in countless technologies today and her example speaks of the American spirit of discovery that exists in every little girl and little boy who know that to look beyond the heavens is to look deep within ourselves, and to figure out just what is possible."

IF THEY WERE NERVOUS IT DIDN'T show, as the men waved goodbye and left Kennedy Space Center en route to Launch Complex 39, where their ride to the moon patiently waited.

2

The Journey. The Landing. The Moon

As the world watched and ground control held its breath, lunar touchdown—and much more—was achieved

Norman Mailer was busy writing 115,000 words on *Apollo 11,* to be serialized over three issues of LIFE magazine and published as a book, *Of a Fire on the Moon,* when he attended a press conference with the three astronauts in 1969. "So now people were going to ask questions of three heroes about their oncoming voyage, which on its face must be in contention for the greatest adventure of man," Mailer wrote. "Yet it all felt as if three young junior executives were announcing their corporation's newest subdivision."

To a man, the astronauts appeared unflappable, largely devoid of poetry or even philosophy, as if they were engineers on a business trip, which indeed they were, as Buzz Aldrin's expense report would later show (see page 96). Neil Armstrong might have carried a briefcase with him on the morning of July 16, 1969, as he emerged from the Manned Spacecraft Building to make the

THE *APOLLO 11* LAUNCH WAS a star-studded event—former President Lyndon Johnson (center) and then Vice President Spiro Agnew (right) stood in the VIP viewing site at Kennedy Space Center. Agnew later said the day signified a "new era in civilization."

AT 9:32 A.M. EASTERN TIME ON July 16, 1969, *Apollo 11* launched into the sunny skies of Florida. After a successful liftoff, ground control (opposite) in Florida breathed a sigh of relief, as (from left) Charles Mathews, Wernher von Braun, George Mueller, and Lieutenant General Samuel Phillips, director of the Apollo program, shared a laugh before it was back to business.

historic walk to *Columbia,* which sat atop the Saturn V rocket on Launch Pad 39A at Cape Kennedy.

"Having taken the elevator up the 320 feet to the level of the waiting spacecraft," writes James R. Hansen in his 2012 Armstrong biography *First Man,* "Armstrong grasped the overhead handrail of the capsule with both hands and swung himself through the hatch. Prior to climbing in, Neil received a small gift from Guenter Wendt, the pad leader: it was a crescent moon that Wendt had carved out of Styrofoam and covered with metal foil. Wendt told him 'it's a key to the Moon,' and a smiling Neil asked Wendt to hold on to the token for him until he got back. In exchange Neil gave Wendt a small card he had been keeping under the wristband of his Omega watch. It was a printed ticket for a ride in a 'space taxi,' reading 'good between any two planets'."

It was in this atmosphere, at 9:32 a.m. on Wednesday, July 16, with as many as a million people watching on the ground—tailgating, drinking, sunbathing on car hoods, selling souvenirs, renting parking lots, idling in boats—that *Apollo 11* trembled. Janet Armstrong and their two boys, Rick, 12, and Mark, six, watched from a friend's yacht on the Banana River, five miles from the launchpad, accompanied by Dodie Hamblin of LIFE, which

Within hours, the astronauts had a full view of the planet they'd just left, the entire spheroid on display outside the capsule window.

A WORLDWIDE AUDIENCE OF HALF a billion people watched the launch on TV, but the view in person was unmatched. People from all over camped out on Florida's beaches (all photos here are from Titusville) to witness history. Said NASA of the weather that day: "Highly suitable . . . winds 10 knots from the southeast, temperature in the mid-80s, and clouds at 15,000 feet."

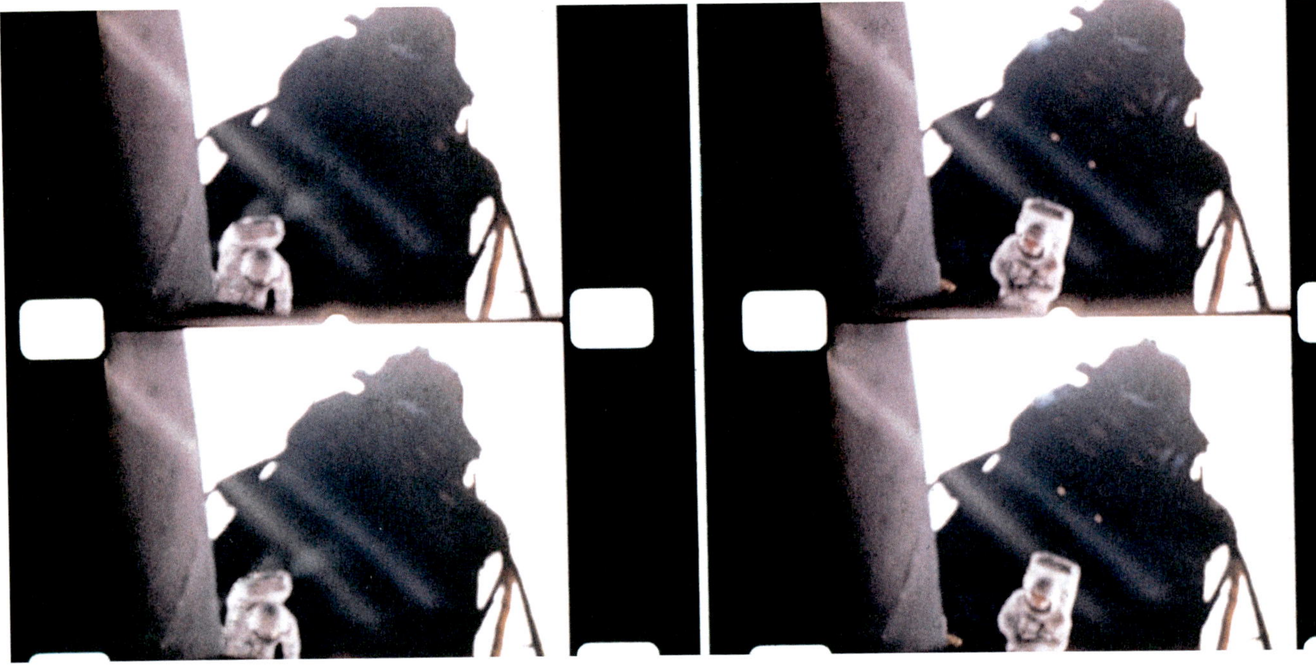

WITH BATED BREATH, ALL corners of the world watched as the *Eagle* landed on the moon and Neil Armstrong descended a ladder, no more excited than if he were climbing down a ladder after painting the side of his house. As he became the first human to ever walk on the moon, he spoke the words we'll never forget: "That's one small step for man, one giant leap for mankind."

had secured exclusive access to the Armstrongs.

Half of the members of Congress, two Supreme Court justices, former President Lyndon Johnson, and sitting Vice President Spiro Agnew were among the eyewitnesses, while President Nixon watched from the White House, after his aborted attempts at that prelaunch dinner. Sargent Shriver, the U.S. ambassador to France, expressed his regret that his deceased brother-in-law, President John F. Kennedy, "isn't alive to see his dream come true." *Tonight Show* host Johnny Carson and his sidekick, Ed McMahon, had flown down from New York. The comedian Jack Benny joined them in the VIP grandstand, filled with famous guests of NASA.

Only the night before, Wernher von Braun, the rocketry pioneer and architect of America's space program, said: "What we will have attained when Neil Armstrong steps down upon the moon is a completely new step in the evolution of man. For the first time, life will leave its planetary cradle, and the ultimate destiny of man will no longer be confined to these familiar continents that we have known so long."

And then the moment was at hand. "T minus 25 seconds. Twenty seconds and counting," intoned public affairs officer Jack King. "T minus 15 seconds, guidance is internal. Twelve, eleven, ten, nine, ignition sequence starts. Six, five, four, three, two, one, zero, all engine[s] running. Liftoff! We have a liftoff. Thirty-two minutes past the hour. Liftoff on *Apollo 11.*" *Apollo 11* rose on its column of fire, "frightening and beautiful all at once," Carson said. Armstrong, Aldrin, and Michael Collins were on their way. Those uncountable eyewitnesses shielded their eyes from the sun as *Apollo 11* and its 48 tons were borne into eternity on fire and steam, and a billion held breaths were exhaled all at once.

Within hours, the astronauts had a full view of the planet they'd just left, the entire spheroid on display outside the capsule window. The crew of *Apollo 11* was struck by the apparent fragility of the earth below. Like a globe on a child's desk, except that this globe lacked the contrived international borders imposed

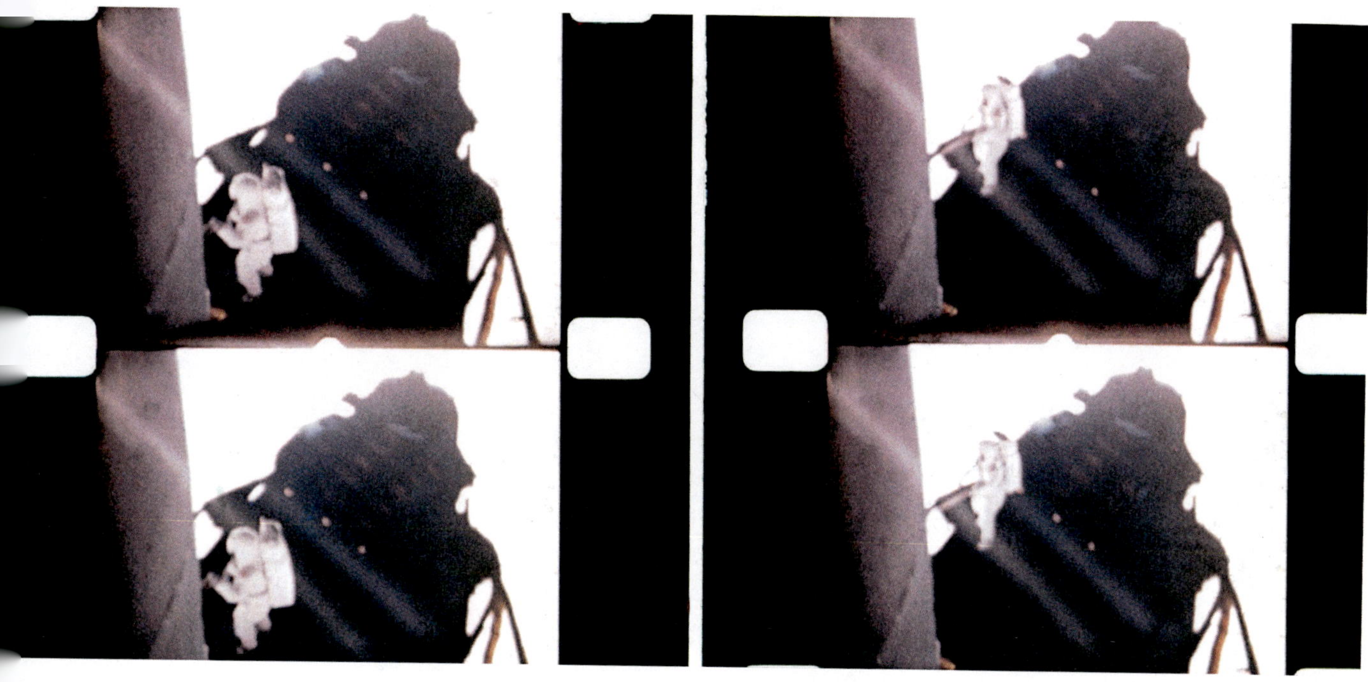

by lines drawn on a map. There were no primary colors to distinguish one country from another, despite the territorial conflicts ranging beneath the cloud cover.

Nature's borders, however, could be made out—rivers, coastlines, mountain ranges. Armstrong, a geography buff, described them to his crewmates. Twenty-five hours and 53 seconds after liftoff, *Apollo 11* passed the midway point of its journey to the moon, slowing to 3,000 miles per hour from a high of 24,200 miles per hour. "It's really a fantastic sight through the sextant," Collins told Houston, as if he were steering by the stars like an ancient mariner. "The reticle just swept across the Mediterranean. You could see all of North Africa, absolutely clear all of Portugal, Spain, southern France. All of Italy absolutely clear. Just a beautiful sight." Replied the voice from Mission Control: "We envy you the view up there."

By then, Janet Armstrong had returned home with her "dead-tired children" in tow. They had flown by private plane from Florida after the launch back to the Houston suburb of El Lago, where she told waiting reporters, "I don't feel historic." Nor could she. Eric, 12, was supposed to play in his Little League baseball all-star game the night his dad left for the moon, but the exhausted boy skipped the game and went to bed.

But it was a reminder that the mundane details of everyday life continued, even in the vacuum of space. And so earthlings were treated to some banter between Collins and Mission Control en route to the moon. "Well it looks like it's about dinner time down there, Earth," Collins said, revealing a food container. "Would you believe you're looking at chicken stew? All you have to do is add three ounces of hot water and blend for five or 10 minutes."

"Sounds delicious," Houston replied.

Among Armstrong's favorite *Apollo 11* astronaut meals were pineapple fruitcake cubes and spaghetti with meat sauce; Aldrin liked the shrimp cocktail. All the quotidian details of space flight—what they ate, how long they slept, precisely how they went to the bathroom—were reported

Beautiful, beautiful. Magnificent desolation.
—BUZZ ALDRIN

LOOKING NORTH OUT OF *APOLLO 11'S* lunar module window, Armstrong and Aldrin, along with the rest of the world, watched as they approached the Sea of Tranquility on Sunday, July 20, eventually touching down at 4:18 p.m. Eastern Time. Armstrong had to maneuver the spacecraft around several boulders on the ground, later saying, "The unknowns were rampant . . . there were just a thousand things to worry about."

in the newspapers. Armstrong chose the music for an audiocassette the astronauts played on *Apollo 11*—Dvorak's *New World Symphony* and a theremin composition called *Music out of the Moon*—and wore an Omega watch. All of these details were a source of endless fascination. As Bowie put it in "Space Oddity": "The papers want to know whose shirts you wear." Every bit of astronaut minutiae devoured on earth was largely ignored by the astronauts themselves, who hadn't had time to read the papers or watch TV as they prepared for their mission. And even their meticulous preparations could not prepare them for everything.

And so it was that *Apollo 11* achieved lunar orbit—leaving the gravitational pull of the earth while feeling the emotional tug of its inhabitants. The archer's arrow neared the target, and the astronauts finally gazed at the immensity of the moon, which hove into view outside the capsule, filling *Columbia* with an almost sinister nimbus.

"The moon I have known all my life, that two-dimensional, small yellow disk in the sky, has gone away somewhere, to be replaced by the most awesome sphere I have ever seen," Collins would write five years later in *Carrying the Fire: An Astronaut's Journeys*. "To begin with it is huge, completely filling our window. Second, it is three-dimensional . . . I almost feel I can reach out and touch it . . . The vague reddish-yellow of the sun's corona, the blanched white of earthshine, and the pure black of the star-studded surrounding sky all combine to cast a bluish glow over the moon. This cool, magnificent sphere hangs there ominously, a formidable presence without sound or motion, issuing us no invitation to invade its domain. Neil sums it up: 'It's a view worth the price of the trip.' And somewhat scary too although no one says that."

"I was worried that the moon might be too soft and that he might sink in," said Armstrong's mom.

Armstrong did, though, in a comment broadcast back to earth, regarding the solar corona, or ring of light: "It looks like an eerie sight," he said.

"It looks very much like the pictures," Armstrong said. "But like the difference between watching the real football game and one on TV. There's no substitute for actually being there."

On Friday afternoon, July 18, Armstrong climbed through the three-foot opening between *Columbia* and the *Eagle* to inspect the vehicle that would or would not land them safely on the lunar surface. "It was the first time they had been inside the landing craft in which

AFTER BOUNDING ABOUT ON the moon's surface for a bit, Armstrong, standing as far from the *Eagle* as either astronaut traveled, snapped a shot of the lunar module at Tranquility Base, 200 feet away at the rim of Little West Crater. Because there's no wind on the moon, the footprint opposite remains intact 50 years later.

they will risk their lives Sunday as they attempt to make the first manned landing on the moon," the *Los Angeles Times* reported on Saturday morning beneath a banner headline—LUNAR LANDER OK—the kind ordinarily reserved for declarations of world war. At the bottom of the page was the usual teaser for the *Times'* astrology column on page 7. This was, after all—as 1969's second most popular (and astrologically minded) song put it—still the Age of Aquarius. For another 48 hours at least, the cosmos remained a source of superstition and soothsaying.

The inspection was part of another TV broadcast back to earth, the astronauts speaking to the world in real time—reality television—and when Armstrong opened the hatch to the LM, a light inside automatically came on. "How about that," said astronaut Charles Duke, the voice of Mission Control in Houston, as if looking at the LM for the first time. "Just like the refrigerator."

These were the astronauts visibly confident, jocular, as giddy as they would ever appear in public. For an instant, their home planet appeared in the capsule window. Armstrong said: "If that's not the earth, we're in trouble."

But they were contriving gallows humor. They had few doubts about the mission at hand. As Collins observed when that moon first filled their capsule window: "I'd just like to get our job done and get out of here."

Before descending to the lunar surface, *Apollo 11* first had to brake, power down, and enter the moon's orbit. Failing to do so would send it on a trajectory back to earth. As the spacemen traveled around the dark side of the moon, they were incommunicado with Houston and the rest of earth for

THE PUBLIC'S OBSESSION with the three astronauts often extended to their families, and the media was desperate for access to them, reporting on the domestic lives of their wives and kids. Janet Armstrong (above) watched the moon landing with her two sons, Mark and Eric; in Texas, Joan Aldrin (opposite, top, in polka-dot blouse) and family watched the moon mission; and Michael Collins's wife, Pat (left, in white), and friends watched from their home in Houston.

23 full minutes. Television viewers breathlessly awaited the radio signal that they were successfully in orbit. When it came—weakly at first, after some suspense and gravely narrated by Walter Cronkite live on CBS—Houston and *Apollo 11* both sounded elated. Said Michael Collins: "Hello, Moon!"

But Collins would admit years later to passing over the cratered landing target and feeling something like the opposite of tranquility. "I don't see any place smooth enough to park a baby buggy," he would later write, "never mind a lunar module."

That Saturday night, before the fateful Sunday when—one way or another—they would both pass into eternity, Armstrong and Aldrin slept little or not at all. They rose at six a.m. Eastern Time on July 20, 1969, while Neil's mother, Viola, back on earth in Ohio, was already in church, praying.

Armstrong and Aldrin's descent, 63 miles from *Columbia* to the Sea of Tranquility in the LM christened the *Eagle,* would leave Collins all by himself in the mother ship. This in itself would be an audacious act, within the larger audacity of flying to the moon. If all went well, the powered descent "would begin with the LM flying horizontally over the moon at a speed of just under 3,800 miles per hour," as Andrew Chaikin put it in *A Man on the Moon,* "and would end with a touchdown as gentle as a leap from a bar stool."

Armstrong and Aldrin stood upright in the seatless *Eagle,* straphangers on the commute that would take the *Eagle* the final 50,000 feet to touchdown on the moon. The *Eagle* separated from *Columbia*—"*Eagle*'s undocked. The *Eagle* has wings"—and proceeded on a flight path the astronauts had dubbed U.S. Highway 1.

As the moon rose up to meet the *Eagle,* the bird was overflying its target, in danger of landing on the slope of a boulder-riddled crater. But Armstrong maneuvered the *Eagle* over a smooth patch of moonscape. "I got a good spot," he told Mission Control. Both men could see the shadow of the *Eagle* on the surface of the moon and could tell that the landing gear was in place.

The world watched, transfixed. At 4:18 p.m. Eastern Time, Mission Control radioed, "We copy you down, *Eagle.*" And Armstrong replied: "Houston, Tranquility Base here. The *Eagle* has landed."

The *Eagle* had 50 seconds of fuel remaining in its descent engine when

CONTINUED ON PAGE 64

"All the News That's Fit to Print"

The New York Times

LATE CITY EDITION

VOL. CXVIII.. No. 40,721 — NEW YORK, MONDAY, JULY 21, 1969 — 10 CENTS

MEN WALK ON MOON

ASTRONAUTS LAND ON PLAIN; COLLECT ROCKS, PLANT FLAG

Voice From Moon: 'Eagle Has Landed'

EAGLE (the lunar module): Houston, Tranquility Base here. The Eagle has landed.

HOUSTON: Roger, Tranquility, we copy you on the ground. You've got a bunch of guys about to turn blue. We're breathing again. Thanks a lot.

TRANQUILITY BASE: Thank you.

HOUSTON: You're looking good here.

TRANQUILITY BASE: A very smooth touchdown.

HOUSTON: Eagle, you are stay for T1. [The first step in the lunar operation.] Over.

TRANQUILITY BASE: Roger. Stay for T1.

HOUSTON: Roger and we see you venting the ox.

TRANQUILITY BASE: Roger.

COLUMBIA (the command and service module): How do you read me?

HOUSTON: Columbia, he has landed Tranquility Base. Eagle is at Tranquility. I read you five by. Over.

COLUMBIA: Yes, I heard the whole thing.

HOUSTON: Well, it's a good show.

COLUMBIA: Fantastic.

TRANQUILITY BASE: I'll second that.

APOLLO CONTROL: The next major stay-no stay will be for the T2 event. That is at 21 minutes 26 seconds after initiation of power descent.

COLUMBIA: Up telemetry command reset to reacquire on high gain.

HOUSTON: Copy. Out.

APOLLO CONTROL: We have an unofficial time for that touchdown of 102 hours, 45 minutes, 42 seconds and we will update that.

HOUSTON: Eagle, you loaded R2 wrong. We want 10254.

TRANQUILITY BASE: Roger. Do you want the horizontal 55 15.2?

HOUSTON: That's affirmative.

APOLLO CONTROL: We're now less than four minutes from our next stay-no stay. It will be for one complete revolution of the command module.

One of the first things that Armstrong and Aldrin will do after getting their next stay-no stay will be to remove their helmets and gloves.

HOUSTON: Eagle, you are stay for T2. Over.

Continued on Page 4, Col. 1

VOYAGE TO THE MOON

By ARCHIBALD MACLEISH

PRESENCE among us,

wanderer in our skies,

dazzle of silver in our leaves and on our waters silver,

A Powdery Surface Is Closely Explored

By JOHN NOBLE WILFORD

HOUSTON, Monday, July 21—Men have landed and walked on the moon.

Two Americans, astronauts of Apollo 11, steered their fragile four-legged lunar module safely and smoothly to the historic landing yesterday at 4:17:40 P.M., Eastern daylight time.

Neil A. Armstrong, the 38-year-old civilian commander, radioed to earth and the mission control room here:

"Houston, Tranquility Base here. The Eagle has landed."

"That's one small step for man, one giant leap for mankind."

Tentative Steps Test Soil

"The surface is fine and powdery," the astronaut reported.

FRONT PAGE NEWS IF EVER there was, the successful moon landing was trumpeted around the world, and Armstrong, Aldrin, and Collins became household names overnight. Now, all they had to do was get home.

OIE TRIOMPHALE POUR MERCKX

Parisien libéré

DE CRÉTEIL A VINCENNES

ÇA Y EST !

S HOMMES LA LUNE !

A 21 h. 17 LE "LEM" A ATTERRI... comme prévu

(malgré la présence du satellite espion soviétique

LA LUNE VAINCUE

A 21 H 17 LE LEM S'EST POSÉ EN DOUCEUR SUR SES QUATRE PATTES

L'AURORE

LUNDI 21 JUILLET 1969

ENCORE UN DRAME CHEZ LES KENNEDY

Le sénateur Edward tue, dans un accident d'auto, la secrétaire de son frère

HIER SUR LE CANAL DE SUEZ

BATAILLE AÉRIENNE et terrestre LA PLUS IMPORTANTE depuis la guerre d...

UONS !

« vainqueur total » du Tour

OUVERTE TOUT L'ÉTÉ

La Librairie Nouvelle

peut expédier dans les 48 heures DES LIVRES POUR VOS VACANCES

Humanité

ORGANE CENTRAL DU PARTI COMMUNISTE FRANÇAIS

...NU DU FOND DES AGES RÉALISÉ

...ME EST SUR LA LUNE

...ET ARMSTRONG SE SONT POSÉS HIER

...1969 à 21 heures 17 minutes 42 secondes

LES DEUX ASTRONAUTES

ALUNISSAGE

É !

PARIS JOUR

Neil Armstrong

Edwin Aldrin

ILS ONT MIS LE PIED SUR LA LUNE

Suite page 24.

PAGES 4, 5 et 6 : Tout sur l'exploit

LE FIGARO

toujours bien glacé CAMPARI à l'eau gazeuse l'apéritif

EDITION DE 5 HEURES

LUNDI 21 JUILLET 1969

...EMIERS HOMMES SUR L...

...A 21 H 17 DANS LA MER DE LA TRANQUILLITÉ

...EAGLE a aluni »; Collins...

LE FRONT DU...

L'aviation...

...MBAT

...OUVEAU MONDE

...SSAGE DU LEM D'APOLLO 11 HIER ...17 ANNONCE UN AGE DIFFERENT

...et Aldrin devraient faire ... pas sur la Lune

INTERNATIONAL Herald Tribune

Published with The New York Times and The Washington Post

PARIS, MONDAY, JULY 21, 1969

MAN ON MOON

Two Astronauts Land Craft Safely, Prepare to Walk on Surface Today

SHOTS OF THE LUNAR module's ascent with a view of earth 238,000 miles away put things into perspective for the astronauts. As Armstrong told a group of high schoolers later that year, "[Earth] was very bright, very beautiful, and very small. It needs to be kept in that beautiful manner."

THE NASA TRANSCRIPTS

Here—from NASA public affairs officer Jack King's counting down of the launch, to the first steps on the moon, to setting course for home—excerpts from communications between Mission Control and Apollo 11

KING: T minus 60 seconds and counting. We've passed T minus 60. 55 seconds and counting. Neil Armstrong just reported back, 'It's been a real smooth countdown.' We've passed the 50-second mark. Power transfer is complete—we're on internal power with the launch vehicle at this time. 40 seconds away from the *Apollo 11* liftoff. All the second-stage tanks now pressurized. 35 seconds and counting. We are still go with *Apollo 11.* 30 seconds and counting. Astronauts report, 'It feels good.' T minus 25 seconds. 20 seconds and counting. T minus 15 seconds, guidance is internal. 12, 11, 10, 9, ignition sequence starts... 6, 5, 4, 3, 2, 1, zero, all engine running... Liftoff! We have a liftoff, 32 minutes past the hour. Liftoff on *Apollo 11*... Tower cleared.

[*At times the voyage to the moon resembled a long family car trip, as when Michael Collins wanted to photograph history's most spectacular sunrise through the window of* Columbia *but couldn't locate his Hasselblad camera, setting off a frantic search of the space capsule.*]

COLLINS: Jesus Christ, look at that horizon!
ARMSTRONG: Isn't that something?
COLLINS: God damn, that's pretty. It's unreal.
ARMSTRONG: Get a picture of that.
COLLINS: Ooh, sure, I will. I've lost a Hasselblad... Has anybody seen a Hasselblad floating by? It couldn't have gone very far—big son of a gun like that... Well, that pisses me off. Hasselblad gone. Find that mother before she or I ends the... Everybody look for a floating Hasselblad. I see a pen floating loose down here, too. Is anybody missing a ballpoint pen?
ALDRIN: Got mine. Is it ballpoint, or is it...?
COLLINS: Yes, ballpoint. Here it is. I mean felt tip.
COLLINS:... Much embarrassed to say [I've] lost a Hasselblad. I seem to be prone to that.
ARMSTRONG: And we're about 7 minutes away, so we got about 7 minutes of practice time.
COLLINS: I've looked—I've looked everywhere over here for that Hasselblad, and I just don't see it...
ARMSTRONG: It's too late for sunrise, anyway.

[*moments later*]

COLLINS: Ah! Here it is!
ARMSTRONG: Find it?
COLLINS: Yes.
ARMSTRONG: Beautiful.
COLLINS: It was floating in the aft bulkhead.

[*A lot of the banter is about taking pictures, in*

this case of the moon.]

ALDRIN: Look at that, would you? Look at that.
COLLINS: Isn't that beautiful?
ARMSTRONG: Pretty good.
COLLINS [*quoting English poet John Keats*]: A thing of beauty is a joy forever.

[*later*]

ARMSTRONG: What are you doing, Mike? What you taking pictures of?
COLLINS: Oh, I don't know. Wasting film, I guess.

[*Onboard alarms sound during the powered descent of the* Eagle—*codes 1201 and 1202. Mission Control in Houston and the astronauts proceed unswayed.*]

HOUSTON: Altitude about 46,000 feet, continuing to descend ... 2 minutes 20 seconds [into the burn of the descent engine]. Everything looking good.
EAGLE: Our position checks downrange here seem to be a little long.
HOUSTON: Eagle, you are go—you are go to continue power descent.
EAGLE: We've got good [radar] lock on. Altitude lights out ... And the earth right out our front window.
EAGLE: 1202, 1202!
HOUSTON: Good radar data. Altitude now 33,500 feet.
EAGLE: Give us a reading on the 1202 program alarm.
HOUSTON: Roger. We got—we're go on that alarm.
HOUSTON: Still go. Altitude 27,000 feet.
EAGLE: [We] throttle down better than in the simulator.
HOUSTON: Altitude now 21,000 feet. Still looking very good. Velocity down now to 1,200 feet per second.
HOUSTON: Eagle, you're looking great, coming up 9 minutes.
HOUSTON: We're now in the approach phase, looking good. Altitude 5,200 feet.
EAGLE: Manual auto altitude control is good.
HOUSTON: Altitude 4,200.
HOUSTON: You're go for landing. Over.
EAGLE: Roger, understand. Go for landing. 3,000 feet.
EAGLE: 12 alarm. 1201.
HOUSTON: Roger, 1201 alarm.
EAGLE: We're go. Hang tight. We're go. 2,000 feet. 47 degrees.
HOUSTON: Eagle looking great. You're go.
HOUSTON: Altitude 1,600 ... 1,400 feet.
EAGLE: 35 degrees. 35 degrees. 750, coming down at 23, 700 feet, 21 down, 33 degrees. 600 feet, down at 19 ... 540 feet ... 400 ... 350 down at 4 ... We're pegged on horizontal velocity. 300 feet down 3½ ... a minute. Got the shadows out there ... altitude-velocity lights. 3½ down, 220 feet, 13 forward, 11 forward, coming down nicely ... 75 feet, things looking good.
HOUSTON: 60 seconds.
EAGLE: Lights on. Down 2½. Forward. Forward. Good. 40 feet down 2½. Picking up some dust. 30 feet, 2½ down. Faint shadow. 4 forward. Drifting to the right a little.
HOUSTON: 30 seconds.
EAGLE: Drifting right. Contact light. Okay, engine stop.
HOUSTON: We copy you down, Eagle.
EAGLE: Houston, Tranquility Base here. The Eagle has landed.
HOUSTON: Roger, Tranquility. We copy you on the ground. You got a bunch of guys about to turn blue. We're breathing again. Thanks a lot.

[*Armstrong begins to back down the ladder of the* Eagle *to the lunar surface.*]

HOUSTON [*Bruce McCandless*]: Okay. Neil, we can see you coming down the ladder now.
ARMSTRONG: Okay. I just checked getting back up to that first step, Buzz. It's ... the strut isn't collapsed too far, but it's adequate to get back up.
HOUSTON: Roger. We copy.
ARMSTRONG: Takes a pretty good little jump [to get back up to that first rung] ...
ARMSTRONG: I'm at the foot of the ladder. The LM footpads are only depressed in the surface about 1 or 2 inches, although the surface appears to be very, very fine-grained, as you get close to it. It's almost like a powder. Ground mass is very fine.
ARMSTRONG: Okay. I'm going to step off the LM now.
ARMSTRONG: That's one small step for man. One giant leap for mankind.

[*Aldrin and Armstrong are reunited with Collins in* Columbia *and are headed home to earth.*]

HOUSTON: How does it feel up there to have some company?
COLLINS: Damn good, I'll tell you.
HOUSTON: I'll bet. I bet, you'd almost be talking to yourself up there after 10 revs or so.
COLLINS: No, no. It's a happy home here. [It's] nice to have company. As a matter of fact, it'd be nice to have a couple of hundred million Americans up here.
HOUSTON: Roger. Well, they were with you in spirit.
COLLINS: Let them see what they're getting for their money.
HOUSTON: Roger. Well, they were with you in spirit anyway, at least that many. We heard on the news today, 11, that last night—yesterday after you made your landing, *New York Times* came out with a—headlines, the largest headlines they've ever used in the history of the newspaper.
COLLINS: Save us a copy.
ARMSTRONG: I'm glad to hear it was fit to print.

CONTINUED FROM PAGE 60

it touched down. (The lunar module had a separate ascent engine that would return it to *Columbia.*)

"Roger, Tranquility," Mission Control in Houston replied. "We copy you on the ground. You got a bunch of guys about to turn blue. We're breathin' again, thanks a lot."

Those "guys," at Command Center in Houston, many in the short-sleeved shirts, narrow ties, and black-framed glasses of the era, were the most visible manifestation of the 400,000 men and women who put the *Eagle* on the moon.

Collins was alone in the capsule then, sitting in the proverbial tin can, far above the world. And he continued in his orbit. In all he would travel alone in the capsule for 21 and a half hours, and for 48 minutes of that time he was completely out of touch. "Far from feeling lonely or abandoned, I feel very much a part of what is taking place on the lunar surface," he would recall 40 years later. "I know that I would be a liar or a fool if I said that I have the best of the three *Apollo 11* seats, but I can say with truth and equanimity that I am perfectly satisfied with the one I have. This venture has been structured for three men, and I consider my third to be as necessary as either of the other two. I don't mean to deny a feeling of solitude. It is there, reinforced by the fact that radio contact with earth abruptly cuts off at the instant I disappear behind the moon. I am alone now, truly alone, and absolutely isolated from any known life. If a count were taken, the score would be three billion plus two over on the other side of the moon, and one plus God knows what on this side."

At Tranquility Base, Armstrong and Aldrin gave their location to Mission Control and to Collins. They took instrument readings and reported on the lunar landscape out their window. In the six hours inside the *Eagle,* parked on the lunar surface, Armstrong thought for the first time about what he might say when taking his first step. Or so he would tell his biographer, James R. Hansen, who wrote: "Armstrong maintains he spent no time thinking about what he would say until sometime after he had successfully executed the landing."

A PLAQUE THAT READ, IN part, "We came in peace for all mankind," signed by President Richard Nixon as well as astronauts Armstrong, Aldrin, and Collins, was left on the moon, as was a solar wind sheet that Aldrin unfurled (opposite) to collect atomic particles from the sun to be sent back to earth.

In those fraught hours before walking on the moon, the astronauts ate a pregame meal of sorts. "I would like to request a few moments of silence," Aldrin said to Mission Control. He had brought with him a vial of wine, a miniature chalice, and a Holy Communion wafer given to him by his Presbyterian pastor in Houston. "I would like to invite each person listening in, wherever or whoever he may be, to contemplate the events of the last few hours," Aldrin said over the radio, "and to give thanks in his own individual way."

By the time Armstrong backed out of the hatchway and onto the ladder of the *Eagle,* on July 21, the world had been bewitched in front of their TV sets for hours. Ricky Armstrong, hand on his forehead in tension, watched with his mom and siblings in a house full of friends and relatives in El Lago. Andy Aldrin awaited his dad's appearance at their own watch party nearby. The Collins family and their viewing party required two television sets. In Wapakoneta, Stephen and Viola

Armstrong watched in their house on Neil Armstrong Drive, the once-quiet street clogged with news trucks.

Neither Collins nor Aldrin were privy to whatever the first words spoken on the moon's surface might end up being. Armstrong never shared with them what he might say. Armstrong's brother, Dean, however, would later claim in a BBC documentary that Neil had shown him the words, printed on a scrap of paper, months before the moon landing, while the two brothers played Risk, that board game of global domination. That was not Neil's recollection, though, and he dismissed countless other reports, as well: that the line was inspired by J.R.R. Tolkien or a children's game called "Mother May I?"

Whatever the case, at 10:56 p.m. Eastern Daylight Time, Armstrong took his great hop of faith off the bottom rung of the *Eagle*'s ladder and said after failing to sink into the lunar surface: "That's one small step for man, one giant leap for mankind."

There was no backup plan should the engine of the *Eagle* not fire. Collins was powerless to rescue his colleagues.

At Disneyland, Mickey Mouse wore an astronaut helmet, and the crowds stopped riding to watch on televisions. In Las Vegas, gamblers paused over baccarat tables. In New York and other cities, people stared at giant screens set up in public squares and paused in unison in airports to gaze in wonder at overhead Zeniths and Motorolas.

In Houston, LIFE had been watching the families watch their loved ones from their homes in upscale suburbs, the astronaut enclaves of El Lago, Nassau Bay, and Timber Cove. Ann Collins waving to her father in *Columbia,* Ricky Armstrong running his palm through his hair as the *Eagle* touched down. When his father finally set foot on the moon, Ricky's mother cried, "That's the big step!"

Aldrin gazed through the window of the *Eagle.* "That looks beautiful from here, Neil," he said, to which Armstrong replied: "It has a stark beauty all its own. It's like much of the high desert of the United States. It's different, but it's very pretty out here." Then Armstrong did what any American boy might do and took the small ring that had been attached to his soil sample bag and threw it across the moonscape. "You can really throw things a long way up here," he said.

Nineteen minutes later, Aldrin joined Armstrong on the lunar surface and remarked on its magnificent desolation. What immediately sprang to mind, he would later write, was "the magnificence of human beings, humanity, Planet Earth, maturing the technologies, imagination, and courage to expand our capabilities beyond the next ocean, to dream about being on the Moon, and then taking advantage of increases in technology and carrying out that dream—achieving that is magnificent testimony to humanity."

The men collected moon dust and moon rocks. They left medals awarded to Yuri Gagarin and Vladimir Komarov, Russian cosmonauts and close friends who had died in separate aeronautics accidents. The medals had been passed along to the *Apollo 11* crew by the cosmonauts' widows. For Armstrong and Aldrin, the tokens were twin reminders of the perils of their own voyage, because getting off the moon was not assured. The mission was hardly over. Kennedy's mandate was not only to put men on the moon but return them "safely to the earth."

"The whole world celebrated our moon landing," Aldrin would later recall. "But we missed the whole thing because we were out of town." They were men at work, with much still to be done.

There was no backup plan should the engine of the *Eagle* not fire. Collins, circling 60-plus miles above the surface, was powerless to rescue his colleagues. Should they be marooned, Aldrin and Armstrong would likely spend their final hours determining what went wrong, for the benefit of future missions. Otherwise, the astronauts carried no suicide pills nor lethal injections in the event of—as the UPI described it, somewhat lugubriously, on the day of their attempted return—"the never-before-encountered horror of being stranded to die on an alien world."

THERE WAS WORK TO DO IN the three hours Armstrong and Aldrin, pictured here, spent on the moon—from measuring wind patterns to collecting soil. NASA wanted to learn as much as they could about the newly discovered land. They had gone all that way, after all.

AT TIMES IT FELT LIKE A movie, watching the men perform these otherworldly acts, and learning of experiences such as Collins (above) piloting himself solo around the dark side of the moon. Seen close-up, the moon was a revelation, with its powdery composition and myriad craters—that's Daedalus Crater opposite—and, thanks to the heroism of the *Apollo 11* crew, we began to better understand the astronomical body that's captivated humans for so long.

"I really wasn't too concerned about the landing," Janet Armstrong told her husband's biographer in *First Man*. "I felt Neil could do that, if at all possible. But, God, you didn't know if that ascent engine was going to fire the next day. If you listened to the TV, as I did later that evening, the drama was on the landing. Well, forget the landing! Are they going to be able to get off there?"

Presidential speechwriter William Safire had composed a message for just such an event. "With no rescue possible, the men would have to bid the world farewell and close down communication preparatory to suicide or starvation," he later wrote. "It would hardly advance the cause of space exploration to force half a billion viewers and listeners to participate in the agony of their demise." Had that happened, Nixon would have told a disconsolate world, "Fate has ordained that the men who went to the moon to explore in peace will stay on the moon to rest in peace. These brave men, Neil Armstrong and Edwin Aldrin, know that there is no hope for their recovery. But they also know that there is hope for mankind in their sacrifice."

Bowie's "Space Oddity" had also imagined an astronaut marooned in space, in this case floating in his "tin can / Far above the world / Planet

Earth is blue / And there's nothing I can do."

But the dreaded nightmare wouldn't come to pass. Nixon would not have to deliver that address to the nation. Twenty-two hours after the *Eagle* landed on the moon, it powered up to rendezvous with Collins in *Columbia,* then 69 miles above. As they rose from the lunar surface, Armstrong and Aldrin left behind magnificent desolation, eternal footprints, and more. "Nine hours after his arrival," LIFE noted, "man had littered the moonscape with his paraphernalia—TV camera, flag, laser reflector, solar wind sheet, special stereo camera, and the sun-powered seismometer."

"The *Eagle* is back in orbit," Armstrong called over the radio, "having left Tranquility Base and leaving behind a replica from our *Apollo 11* patch and the olive branch." They also left behind, of course, the descent stage of the *Eagle,* forever marooned on the Sea of Tranquility.

"Roger, we copy," replied astronaut Ron Evans at Mission Control in Houston. "The whole world is proud of you."

The *Eagle* began its trip back home at 1:54 p.m. Eastern Time and reunited with Collins and *Columbia* at 5:35 p.m. Asked by Houston how it felt to have company again, Collins answered: "Damn good, I'll tell you."

A little more than two hours after that rendezvous, the ascent stage of the *Eagle* was jettisoned. Fifty years later, NASA would list the whereabouts of that historic vehicle as "impact site unknown."

And thus began their journey home, during which the astronauts each enjoyed extended and well-earned

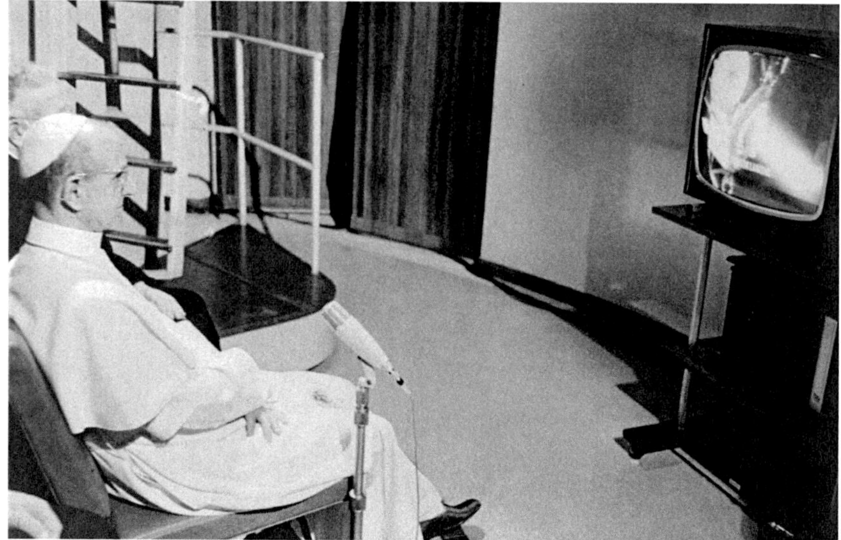

ON FEW OCCASIONS DOES THE Holy Father drop what he's doing to flip on the tube, but that's what Pope Paul VI (opposite, bottom) did to watch the guys touch down. Meanwhile, Muscovites (opposite, top) read the July 21 edition of the Soviet newspaper *Izvestia* to learn the news. TV news anchor Walter Cronkite (above) was glued to the screen, as was a Japanese family in Tokyo (above, right) along with crowds at JFK airport in New York City, pictured here.

The astronauts returned to their home planet as Lindbergh had landed at Le Bourget airfield in Paris 42 years earlier—the objects of global obsession.

periods of sleep. As they did on the outbound leg of their journey, they took photographs and appeared on television, but this time the earth's gravitational pull was inexorably drawing them home.

The world was still watching, of course, and still reacting to Sunday's achievement. South Korea named its first superhighway, between Seoul and Inchon, Apollo. The British bookmaking behemoth William Hill offered 100-to-1 odds that man would land on Mars by July 20, 1976—and even-up odds for July 20, 1979.

The astronauts' return to earth—a phrase that now evokes anticlimax—was not exactly uneventful. This voyage home was, after all, another first in human history. But if it looked from earth like a victory lap for the Apollo program, the men in the capsule remained humble, casual, and calm.

"So what's new?" Collins asked Bruce McCandless in Houston while hurtling toward earth.

"Oh, we were wondering what's new with you up there?" McCandless replied.

"All very quiet," Collins said. And then, after a brief description of the thruster damping down, the astronaut said, "Nice to sit here and watch the earth getting larger and larger, and the moon getting smaller and smaller."

By the time the command module made its fiery reentry into the earth's atmosphere, as lunchtime approached in Houston on July 24, the capsule appeared to be combusting. But the flames caused by the friction of reentry were in fact burning up the protective ablative material on the ship. The capsule itself was safe, as were its three world-famous explorers, about to embark on a new journey—of ticker tape, presidential palaces, and outsize international fame.

But first, three red-and-white-striped parachutes bloomed, softening *Columbia*'s descent. They resembled sidewalk cafe umbrellas—jaunty, celebratory. Mrs. Neil Armstrong knelt in prayer in front of her TV in El Lago as the three men splashed down in the Pacific Ocean. She wasn't going to pop champagne until her husband and his co-workers were safely returned from the moon. The three men had made a round trip of half a million miles and came back to earth within a mile of their targeted landing spot, 812 nautical miles southwest of Hawaii, and just 12 miles from the waiting USS *Hornet*.

The astronauts returned to their home planet as Lindbergh had landed at Le Bourget airfield in Paris 42 years earlier—the objects of global obsession. Asked before the launch if he was prepared to surrender his privacy for the good of the mission, Armstrong replied, "I think that's a fair trade." He was about to find out.

It was 5:50 a.m. local time, and *Columbia* capsized, overturned by waves, but airbags deployed and returned the vessel upright. It bobbed like a bath toy. Indeed, smaller versions of this spacecraft would become bath toys. A Navy helicopter dropped frogman John Wolfram, who attached

an anchor to *Columbia.* Rafts were dropped and inflated. A rubber collar was attached to *Columbia* to stop its violent bobbing, lest motion sickness overcome these moon men in the final seconds of their journey. But the real fear in retrieving the astronauts from the ocean was not seasickness. It was lunar-borne disease.

The three astronauts emerged through the capsule hatch and were immediately handed "biological isolation garments"—BIGs—to wear en route to quarantine. They boarded the raft and were washed down with a chemical solution, hoisted up to the hovering helicopter, and landed on the USS *Hornet,* where they were placed in a quarantine trailer 68 minutes after splashdown, sealed inside like money in a vault, with little to keep them occupied beyond books and a smuggled ukulele. Michael Collins sported the first mustache cultivated in space.

AFTER A TRIP OF 195 HOURS, 18 minutes, and 35 seconds (36 minutes longer than intended), the three astronauts splashed down in the Pacific Ocean 812 nautical miles southwest of Hawaii and were pulled from the *Apollo 11* command module with the help of frogmen who transported them via helicopter to the deck of the USS *Hornet*. All were required to wear biological isolation garments.

To see if men who had traveled to the moon were contagious to their fellow earthlings, laboratory mice were to be injected with moon dust. "Their reactions, the findings of autopsies when they are killed, and tests on human and animal cells and 33 types of plants will determine whether the moon is safe and the astronauts can go home," reported LIFE. "Or whether they have opened a Pandora's box of unearthly medical problems."

The astronauts instead encountered a well of happiness, a triumph impossible to be undone. Inoculated against everything but their white-hot international fame, the men spoke to Nixon by microphone through a window display in their quarantine trailer aboard the *Hornet*. No one thought it hyperbolic when the President told the astronauts through the glass: "This is the greatest week in the history of the world since the creation."

Pope Paul VI, 71, who watched the moon landing from the papal observatory at the pontiff's summer retreat south of Rome, said, "Blessings to you, conquerors of the moon, pale lamp of our nights and our dreams."

For the astronauts, coming down to earth was both a reality and a metaphor. The three men signed customs forms for the moon rocks they brought back, souvenirs of their voyage. They had returned from 238,855 miles away, on an eight-day business trip for which Aldrin would submit an expense report of $33.31, mostly in reimbursement for his personal ground transportation. Under "Points of Travel" on his expense voucher, he typed: "Houston, Texas [to] Cape Kennedy, Fla., Moon, Pacific Ocean (USN Hornett), Hawaii and return to Houston, Texas." ●

A WAVE OF JOY—AND, LIKELY, relief—washed over the flight controllers in the Manned Spacecraft Center after the completion of the voyage of *Apollo 11*, and they let their patriotism fly unabashed.

IN CASE OF DISASTER

Now in the National Archives, this memo written by William Safire contains the speech President Nixon would have delivered to the nation in the event that Armstrong and Aldrin perished on the moon

To : H. R. Haldeman

From: Bill Safire July 18, 1969.

IN EVENT OF MOON DISASTER:

Fate has ordained that the men who went to the moon to explore in peace will stay on the moon to rest in peace.

These brave men, Neil Armstrong and Edwin Aldrin, know that there is no hope for their recovery. But they also know that there is hope for mankind in their sacrifice.

These two men are laying down their lives in mankind's most noble goal: the search for truth and understanding.

They will be mourned by their families and friends; they will be mourned by their nation; they will be mourned by the people of the world; they will be mourned by a Mother Earth that dared send two of her sons into the unknown.

In their exploration, they stirred the people of the world to feel as one; in their sacrifice, they bind more tightly the brotherhood of man.

In ancient days, men looked at stars and saw their heroes in the constellations. In modern times, we do much the same, but our heroes are epic men of flesh and blood.

Others will follow, and surely find their way home. Man's search will not be denied. But these men were the first, and they will remain the foremost in our hearts.

For every human being who looks up at the moon in the nights to come will know that there is some corner of another world that is forever mankind.

PRIOR TO THE PRESIDENT'S STATEMENT:

The President should telephone each of the widows-to-be.

AFTER THE PRESIDENT'S STATEMENT, AT THE POINT WHEN NASA ENDS COMMUNICATIONS WITH THE MEN:

A clergyman should adopt the same procedure as a burial at sea, commending their souls to "the deepest of the deep," concluding with the Lord's Prayer.

To Infinity and Beyond

A half century after that extraordinary mission, the power of the moon landing, and the ideas and explorations it bequeathed, continues to have a special hold

AT 112,000 MILES FROM earth, the *Apollo 11* astronauts had a striking view of Europe, Africa, and the Arabian Peninsula. "Just a beautiful sight," Collins said. Replied Houston: "We envy you the view up there."

PRESIDENT RICHARD NIXON greeted the astronauts in the mobile quarantine facility on the USS *Hornet* upon their return to earth.

They emerged from quarantine, free of lunar maladies, and were embraced by the nation, first in an August ticker-tape parade in New York City, then on to Chicago that very afternoon, and to Los Angeles in the evening, for a dinner at the Century Plaza Hotel hosted by President Nixon and attended by Hollywood's A-list. This transcontinental journey proved grueling even for men who had traveled to the moon and back.

In September, Neil Armstrong returned to Wapakoneta and told a crowd of teenagers gathered at the Wapakoneta High School football field, "Some of you have a favorite song these days which has become a favorite of mine—the dawning of the Age of Aquarius. I believe the age of Aquarius will come." He had steered by the stars, he said, an ancient explorer on a new ocean. "The stars of Aquarius told me the rendezvous point of *Apollo 11*," he said. And he urged his audience to take care of

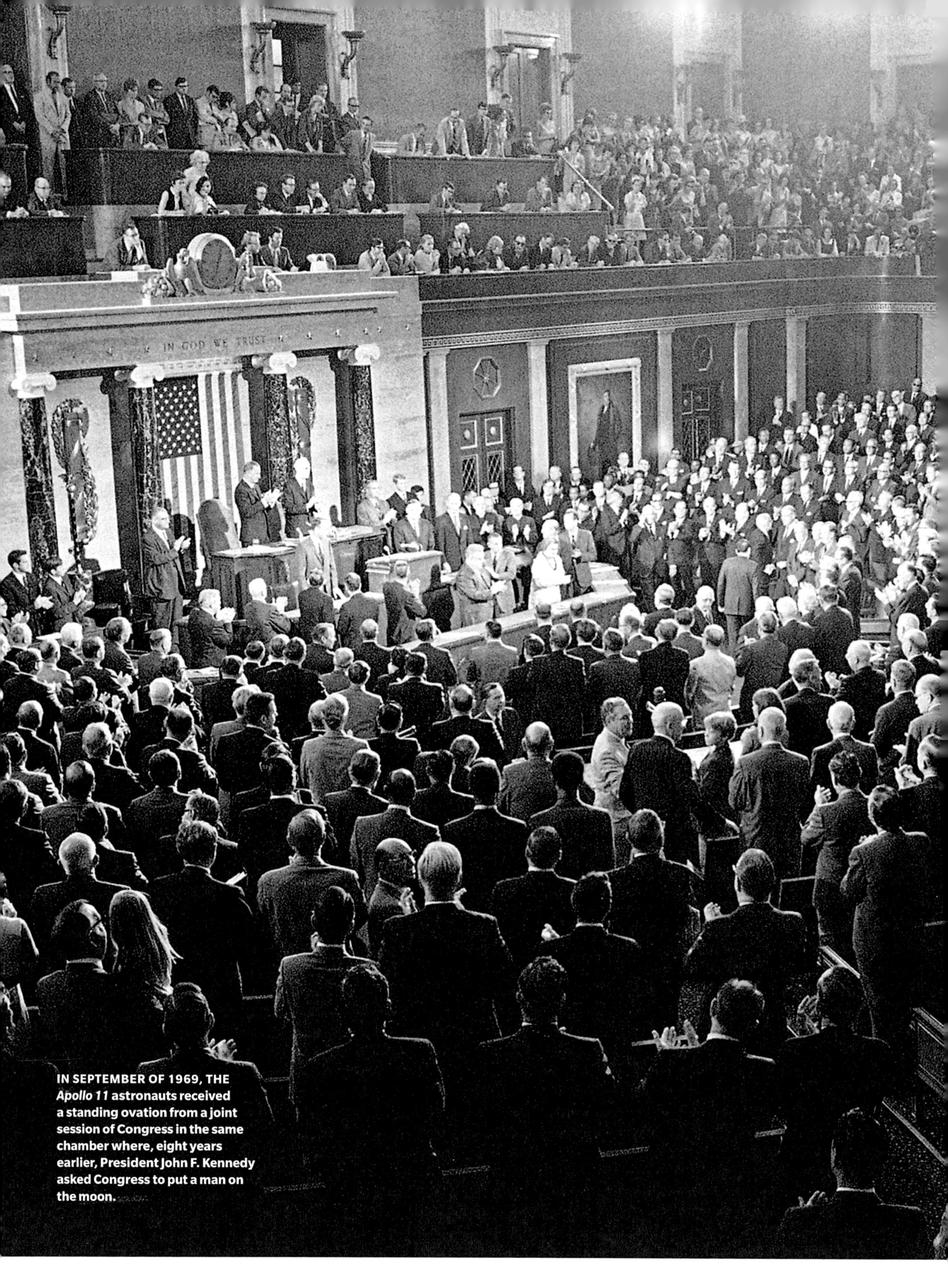

IN SEPTEMBER OF 1969, THE *Apollo 11* astronauts received a standing ovation from a joint session of Congress in the same chamber where, eight years earlier, President John F. Kennedy asked Congress to put a man on the moon.

their home. "As I stood on the moon, I could look up and look at the planet Earth," Armstrong said. "It was very bright, very beautiful and very small. It needs to be kept in that beautiful manner."

Earth's fragility and parochialism would be a recurring theme of the astronauts' public appearances. In October, Armstrong, Buzz Aldrin, and Michael Collins went to London, to meet Queen Elizabeth and Prince Philip at Buckingham Palace. In the span of a month, they would meet nearly every major world leader, including Pope Paul VI, the Shah of Iran, Generalissimo Francisco Franco of Spain, and the Japanese emperor Hirohito in Tokyo, where 120,000 lined the streets to wave. "From space it is impossible to see national borders," Collins told the crowd that greeted them at the airport. "We are all passengers on a voyage in orbit about the sun."

They belonged to the world now, to high culture and low. The American poet Archibald Macleish and the English American poet W.H. Auden wrote pieces about the lunar mission ("Voyage to the Moon" and "Moon Landing," respectively).

Another group of English poets—the rock band Jethro Tull—released a song called "For Michael Collins, Jeffrey, and Me," which goes:

> I'm with you L.E.M
> Though it's a shame that it had to
> be you
> The mother ship
> Is just a blip from your trip made
> for two
> I'm with you boys
> So please employ just a little extra
> care

CONTINUED ON PAGE 87

THE PARTY STARTED IN AUGUST and didn't end for the rest of the year, the first stop a ticker-tape parade through New York City before jetting on to Chicago and then Los Angeles that same day.

PRESIDENT NIXON FETED the *Apollo 11* astronauts (opposite, top), in Los Angeles at the Century Plaza Hotel in August 1969. On June 15, 1970, Pope Paul VI (bottom) greeted the men in the Papal Library at the Vatican. That same year Queen Elizabeth II (above) welcomed the astronauts to Buckingham Palace.

CONTINUED FROM PAGE 83

It's on my mind
I'm left behind when I should
have been there
Walking with you

The astronauts' reward for their cosmic journey was a terrestrial tour of 24 nations. The solitude of space had given way to the opposite on earth. Armstrong traveled with comedian Bob Hope to Thailand and Vietnam, where they appeared before 13,000 soldiers at Lai Khe four days before Christmas. The unlikely comedy duo performed topical bits that sought to diminish the magnitude and danger of what the astronauts had accomplished, out of respect for the risks faced by their audience of GIs, seated on ammunition cases. And so Armstrong's stepping on the moon reminded Hope of the LSD guru "Timothy Leary running for a bus. You really were floating." Armstrong said that his most harrowing moment on *Apollo 11* was "when the door to the washroom jammed."

Hope: "Your step was the second most dangerous step of the year."

Armstrong: "Who took the first most dangerous step?"

Hope: "The girl who married Tiny Tim."

And yet the First Man hadn't forgotten the insights gained 238,000 miles above earth. As he said to the troops: "It is too bad men can't learn together in peace down here."

To be sure, something—some mystery or romance—was lost when Armstrong set foot on the moon and failed to sink into a mysterious substance of one sort or another. "The moon our grandfathers saw at its unimaginable distance in the sky had far more impact—often as a direct result of the fallacy, naivete, and astonishment with which it was regarded—than the new one is ever likely to exert," lamented LIFE magazine. "It inspired fertility rites among Asians and human sacrifice among Celts. It provided goddesses for Babylonians and Egyptians. It told Winnebagos when to plant corn. It stopped the sap in Cuban trees and started it in Broadway song writers. Whadda we got now? Craters, old boy. Craters."

Over the next three years and five months, 10 more men followed Armstrong and Aldrin to the surface of the moon. Pete Conrad and Alan Bean, in *Apollo 12,* set down on the lunar surface on November 19, 1969, four months after their colleagues first landed. James Lovell, Jack Swigert, and Fred Haise famously aborted their *Apollo 13* mission after an oxygen tank exploded. In 1971, Alan Shepard and Edgar Mitchell walked on the moon on the *Apollo 14* mission, and Shepard hit a golf ball on the lunar surface, a drive that likely traveled for miles. David Scott and James Irwin walked on the moon that same year, as *Apollo 15* astronauts. And in 1972 John Young and Charles Duke walked on the moon as part of the *Apollo 16* crew, months ahead of Eugene Cernan and Harrison Schmitt of *Apollo 17*.

Cernan would draw in the moon dust the initials of his then nine-year-old daughter, Tracy. The initials are still there, along with Armstrong's footprint and the American flag and Alan Shepard's golf ball. No man has set foot on the moon since December 14, 1972. But they did leave the light on.

Fifty years later, the moon remains devoid of Hertz lots and Hilton hotels. NASA's marquee program became the space shuttle by the 1980s—when the moonwalk survived principally as the signature dance move of Michael Jackson—and then the International Space Station became the setting for men and women traveling in space. The

NEIL ARMSTRONG GREETED 20,000 servicemen in Long Binh, Vietnam, in December 1969, as part of Bob Hope's annual holiday armed forces tour. Armstrong, an unlikely comedy partner for Hope, told the audience his most harrowing moment aboard *Apollo 11* was when the "door to the washroom jammed."

MP
MP
MP
MP
MP

THE *APOLLO 11* CREW, donning sombreros and ponchos, were swarmed by throngs of admirers in the streets of Mexico City in September 1969 (above); six weeks later, in Sydney, Australia, they were given a parade that proved to be a bit more tame.

Mars Rover is the robotic embodiment of a larger desire to put a man on the red planet.

Billionaire entrepreneurs—Elon Musk and Jeff Bezos—want to take private citizens into space, much as Pan Am and TWA sought to do in the 1960s. The President of the United States wants to establish a new branch of the military called the Space Force. But the moon—the "sacred moon," as Walt Whitman called it, is no longer a priority. Henry David Thoreau's "mistress of the night," as a destination for man, has been largely reduced to that most jaded of traveler's complaints: Been there, done that.

And yet the achievement of the first men who went there has never dimmed, even as the men themselves retreated into private life. Michael Collins became the director of the National Air & Space Museum, part of the Smithsonian Institution in Washington, D.C., where 3,500 items from *Apollo 11* are housed—everything from the command module *Columbia* to Collins's space toothbrush to a pineapple fruitcake that has traveled to the moon and back and remains—after 50 years—uneaten and inedible.

Reentry was more difficult for Buzz Aldrin, who suffered from

ZSF·170

Billionaire entrepreneurs want to take private citizens into space—yet the achievement of the first men who went to the moon has never dimmed.

alcohol dependence and depression—he described his personal life as one of "magnificent desolation"—though he is now four decades sober. The Aldrin Crater on the moon, 50 kilometers from the Sea of Tranquility, bears his name, but Buzz Aldrin is more famously memorialized by the character Buzz Lightyear, the astronaut in the *Toy Story* movie franchise, a space-visored hero whose catchphrase exudes the boundless possibility captured by *Apollo 11*: "To infinity and beyond!"

Lightyear's first name is a source of bemusement to the original Buzz, who has noted that he never got a penny from the filmmakers. Indeed, his most famous moment on video since the moon landing was shot outside a hotel in Beverly Hills, where the astronaut was confronted and verbally provoked by a conspiracy theorist who claimed that the moon landing was a hoax. Aldrin punched the man in the jaw.

"I have heard Frank Sinatra sing 'Fly Me to the Moon' almost too many times," he told fans in a 2014 interview on Reddit. "So I'm interested in composing a new song entitled 'Get Your Ass to Mars!'" He said he had no doubt that humankind's next great achievement would be landing a person on the red planet.

Neil Armstrong lived a more private, less tumultuous afterlife than Aldrin did. In the decade immediately following the moon landing, Armstrong taught aeronautical engineering at the University of Cincinnati. He had no name on his office door, but signed autographs when asked. (Armstrong would stop signing autographs in 1994.) He appeared in commercials for Chrysler, but otherwise Armstrong successfully returned to private life. In one of his final major public appearances, speaking in 2010 at Senate hearings on the future of U.S. human space flight, he expressed his skepticism of privately funded excursions to the cosmos. "I support the encouragement of newcomers toward their

NASA LIVES ON, LOOKING TO the next space frontier to be conquered. Clockwise from top left: In April 2019, a SpaceX Falcon Heavy rocket, carrying the Arabsat 6A communications satellite, lifted off from the Kennedy Space Center; in December 2017, the Blue Origin New Shepard M7 launched; on June 15, 2018, NASA's Curiosity Mars Rover snapped a few selfies in the red planet's Gale Crater; and in March 2019, astronauts Anne McClain (left), Christina Koch, and Nick Hague made sure their suits fit in preparation for some spacewalking.

BLUE ORIGIN

goal of lower-cost access to space," Armstrong said, "but having cut my teeth in rockets more than 50 years ago, I am not confident." He knew the cost (more than $25 billion for the Apollo program) and technical staff (an estimated 400,000) required to get him and his fellow astronauts to the moon and back.

In 2018, Ryan Gosling portrayed Neil Armstrong in *First Man*, the feature film based on James Hansen's best-selling biography. Armstrong did not live to see the film. He died at age 82 on August 25, 2012, two weeks after undergoing quadruple bypass surgery in suburban Cincinnati, not far from where he was born, which was not far from the Wright brothers' bicycle shop.

Armstrong was mourned throughout the world. "Neil was among the greatest of American heroes—not just of his time, but of all time," President Obama said. "When Neil stepped foot on the surface of the moon for the first time, he delivered a moment of human achievement that will never be forgotten."

The astronaut's ashes were transported on the aircraft carrier USS *Philippine Sea* and scattered in the Atlantic Ocean. His spirit also remains in another sea, the Sea of Tranquility, as Armstrong's family acknowledged in a statement after his death. "The next time you walk outside on a clear night and see the moon smiling down at you, think of Neil Armstrong and give him a wink," the family stated. Or as Bing Crosby and Frank Sinatra and Billie Holiday put it, on behalf of their fellow wonder-struck earthlings:

"When the night is new, I'll be looking at the moon, but I'll be seeing you." ●

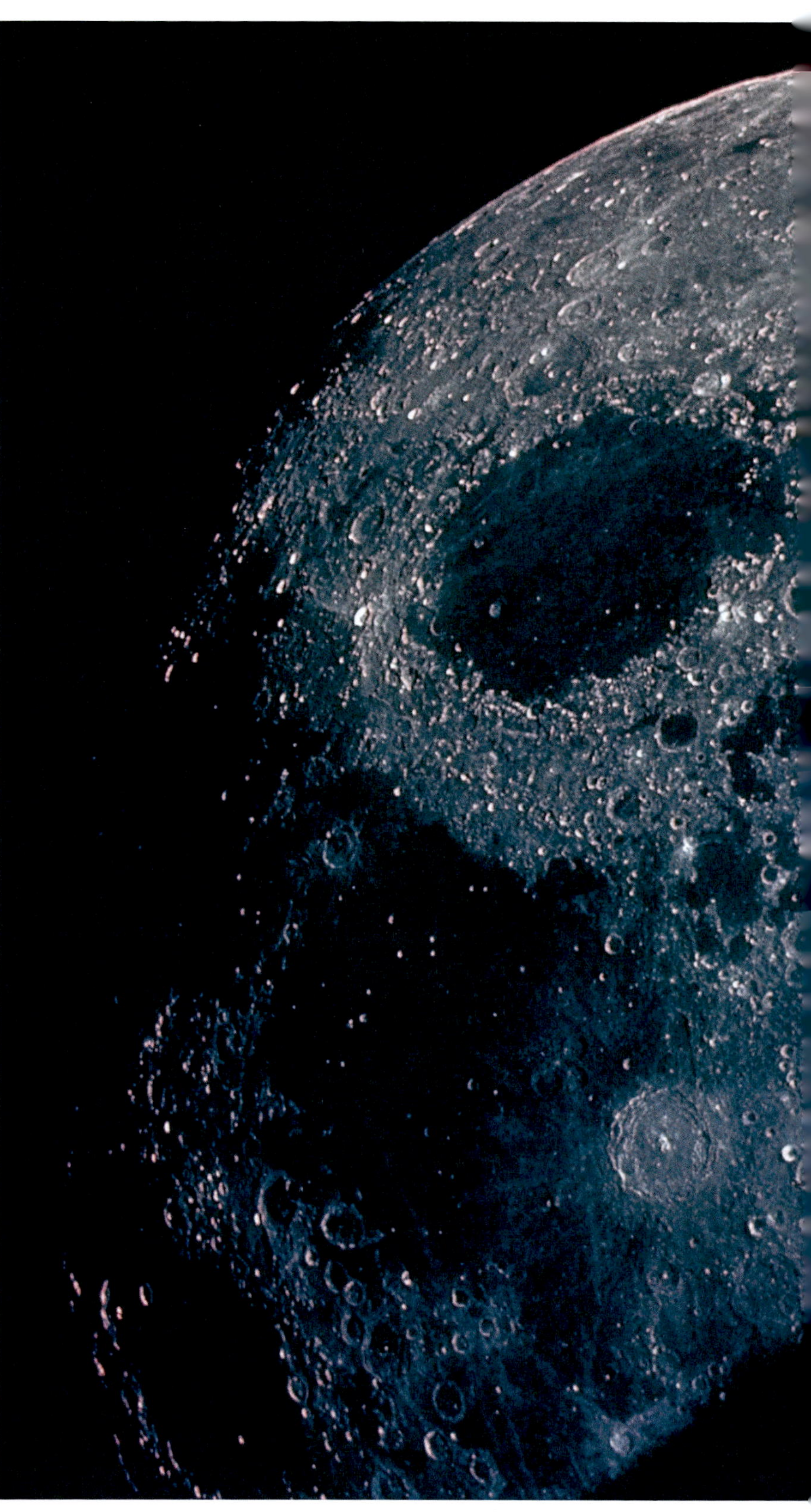

WE HAVEN'T BEEN BACK TO THE moon since the Apollo program ended in 1972, and it doesn't appear we'll return anytime soon. Perhaps that's fine. We took a giant leap in 1969 and, half a century later, it remains one of mankind's finest hours.

PHOTO CREDITS

COVER NASA/Ullstein Bild/Getty
BACK COVER Ralph Crane/LIFE/The Picture Collection
PAGE 1: NASA/Getty
PAGES 2–3: REX/Shutterstock

INTRODUCTION: THE END OF THE IMPOSSIBLE
PAGE 4: (clockwise from top): Apic/Moviepix/Getty; © Look and Learn/Private Collection/Bridgeman Images (2) **PAGE 7:** (clockwise from top left) Ralph Morse/LIFE/The Picture Collection; © The Hospitality Industry Archives, Conrad Hilton College, University of Houston (2); Michael Ochs Archives/Getty; no credit **PAGE 8:** (clockwise from top): Fritz Goro/LIFE/The Picture Collection; Gamma-Keystone/Getty; Buyenlarge/Archive/Getty

BEFORE THE MOON: A STORY IN PICTURES
PAGES 10–11: Ralph Morse/LIFE/The Picture Collection (2) **PAGES 12–13:** NASA **PAGES 14–15:** SSPL/NASA/Getty **PAGES 16–17:** Ralph Morse/LIFE/The Picture Collection

CHAPTER 1: "IS YOUR SON GOING TO FLY OUT INTO SPACE?"
PAGES 19–21: Bettmann/Getty (2) **PAGES 22–23:** Corbis/Getty **PAGE 24:** (from top) Neil Armstrong™ Used with Permission from The Purdue Research Foundation; no credit **PAGE 25:** NASA **PAGES 26–27:** (from left) Seth Poppel/Yearbook Library; Ralph Morse/LIFE/The Picture Collection **PAGE 28:** AP/REX/Shutterstock **PAGE 29:** United States Air Force **PAGE 30:** SSPL/NASA/Getty **PAGE 31:** Robert F. Sisson & Donald McBain/National Geographic/Getty **PAGES 32–34:** Ralph Morse/LIFE/The Picture Collection (2) **PAGE 35:** NASA/The Picture Collection, Inc. **PAGE 36:** NASA/The New York Times/Redux **PAGE 37:** SSPL/NASA/Getty **PAGES 38–40:** Bettmann/Getty (2) **PAGE 41:** O. Louis Mazzatenta/National Geographic Image Collection RF/Getty **PAGE 42:** Courtesy MIT Museum

CHAPTER 2: THE JOURNEY. THE LANDING. THE MOON
PAGES 44–47: NASA (2) **PAGE 48:** Ralph Morse/LIFE/The Picture Collection **PAGE 49:** NASA/The New York Times/Redux **PAGES 50–51:** David Burnett/Contact Press Images (4) **PAGES 52–53:** NASA/The LIFE Picture Collection/Getty **PAGES 54–55:** NASA/The LIFE Picture Collection/Getty **PAGE 56:** NASA/Ullstein Bild/Getty **PAGE 57:** NASA **PAGE 58:** Vernon Merritt III/LIFE/The Picture Collection **PAGE 59:** (from top) Lee Balterman/LIFE/The Picture Collection; Bob Peterson/The LIFE Images Collection/Getty **PAGE 60:** Granger **PAGE 61:** Bettmann/Getty **PAGE 62:** NASA/The Picture Collection, Inc. **PAGE 64:** Michael Ochs Archives/Getty **PAGES 65–67:** NASA/The Picture Collection, Inc. (2) **PAGE 68:** Bob Rice/Fairfax Media/Getty **PAGE 69:** NASA/Science Source **PAGE 70:** Bettmann/Getty (2) **PAGE 71:** (clockwise from bottom) CBS Photo Archive/Getty (2); REX/Shutterstock **PAGES 72–73:** Bettmann/Getty **PAGES 74–75:** NASA **PAGES 76–77:** Courtesy Richard Nixon Presidential Library and Museum (2)

CHAPTER 3: TO INFINITY AND BEYOND
PAGE 79: NASA/Science Source **PAGES 80–81:** MPI/Archive/Getty **PAGES 82–83:** AP/REX/Shutterstock **PAGES 84–85:** Mario De Biasi/Mondadori Portfolio/Getty **PAGE 86:** (from top) ABC Photo Archives/Getty; REX/Shutterstock **PAGE 87:** Hulton/Getty **PAGES 88–89:** Bettmann/Getty **PAGE 90:** REX/Shutterstock **PAGE 91:** George Lipman/The Sydney Morning Herald/Fairfax Media/Getty **PAGES 92–93:** (clockwise from top left) Joe Skipper/Reuters; Blue Origin; AP/REX/Shutterstock; NASA/Reuters **PAGES 94–95:** NASA/The Picture Collection, Inc.

FINAL BILL
PAGE 96: Courtesy Buzz Aldrin

Final Bill

STANDARD FORM 1012-A
Title 7, GAO Manual
1012-210

TRAVEL VOUCHER
MEMORANDUM

DEPARTMENT, BUREAU, OR ESTABLISHMENT
NASA - Manned Spacecraft Center

VOUCHER NO. 014501

PAYEE'S NAME
Col. Edwin E. Aldrin 00018

SCHEDULE NO.

MAILING ADDRESS PLEASE MAKE CHECK PAYABLE TO:
Nassau Bay National Bank
P.O. Box 58008
Houston, Texas 77032 Account #1-0348-9

PAID BY

OFFICIAL DUTY STATION
Houston, Texas

RESIDENCE

FOR TRAVEL AND OTHER EXPENSES
FROM (DATE) 7-7-69 TO (DATE) 7-27-69

APPLICABLE TRAVEL AUTHORIZATION(S)
NO. X-22002 DATE 6/18/69

TRAVEL ADVANCE
Outstanding $
Amount to be applied
Balance to remain outstanding $

CHECK NO.

CASH PAYMENT OF $
RECEIVED (DATE)

TRANSPORTATION REQUESTS ISSUED

TRANSPORTATION REQUEST NUMBER	AGENT'S VALUATION OF TICKET	INITIALS OF CARRIER ISSUING TICKET	MODE, CLASS OF SERVICE, AND ACCOM-MODATIONS *	DATE ISSUED	POINTS OF TRAVEL FROM–	TO–
Gov. Air					Houston, Texas	Cape Kennedy, Fla. Moon Pacific Ocean (USN Hornett) Hawaii
					and return to Houston, Texas	

No Cash

8-4-69

AMOUNT CLAIMED → Dollars 33 Cts 31

APPROVED *(Supervisory and other approvals when required)*

DIFFERENCES:

NEXT PREVIOUS VOUCHER PAID UNDER SAME TRAVEL AUTHORITY
VOUCHER NO. | D.O. SYMBOL | DATE (MONTH–YEAR)

Total verified correct for charge to appropriation(s) (initials) 33 31

Applied to travel advance (appropriation symbol)

AUG 26 1969

C. W. Bird
Authorized Certifying Officer

NET TO TRAVELER → 33 31

ACCOUNTING CLASSIFICATION
039-00-00-00-CA-2031-CB11

* Abbreviations for Pullman accommodations: MR, master room; DR, drawing room; CP, compartment; BR, bedroom; DSR, duplex single room; RM, roomette; DRM, duplex roomette; SOS, single occupancy section; LB, lower berth; UB, upper berth; LB-UB, lower and upper berth; S, seat.

For his round trip to the moon, Aldrin submitted to NASA a $33.31 expense report.

Made in the USA
San Bernardino, CA
18 August 2019